D0959209

DWARF CICHLIDS

DR. JÖRG VIERKE

Title page photo: *Pelvicachromis pulcher.*

Photography: Dr. Herbert R. Axelrod; Heiko Bleher; P. de Rahm; Dr. S. Frank; Dr. Harry Grier; H. Hansen; J. Hansen; R. Hal Holden; H. Jung; B. Kahl; K. Knaack; Horst Linke; H. J. Mayland; Midori Shobo, Japan; Aaron Norman; H. J. Richter; E. Roloff; A. Roth; Harald Schultz; Dr. W. Staeck; R. Stawikowski; Dr. J. Vierke; G. Wolfsheimer; R. Zukal.

Originally published in the German language by Franckh'sche Verlagshandlung, W. Keller & Co., Stuttgart, Germany. The original title was Zwergbuntbarsche im Aquarium: Ihre Pflege und Zucht. The first German edition was copyrighted in 1977.

The first American edition was published in 1979 with considerable new material added to the first German edition.

This is the second English edition, published in 1988. The entire text has been re-edited. The type has been completely re-cast and the book has been re-designed in a more modern format. There are almost twice as many color photographs in this new edition as any previous edition...and there are 32 more pages of text.

Translated by Christa Ahrens

Distributed in the UNITED STATES by T.F.H. Publications, Inc., One T.F.H. Plaza, Neptune City, NJ 07753; in CANADA to the Pet Trade by H & L Pet Supplies Inc., 27 Kingston Crescent, Kitchener, Ontario N2B 2T6; Rolf C. Hagen Ltd., 3225 Sartelon Street, Montreal 382 Quebec; in CANADA to the Book Trade by Macmillan of Canada (A Division of Canada Publishing Corporation), 164 Commander Boulevard, Agincourt, Ontario M1S 3C7; in ENGLAND by T.F.H. Publications Limited, Cliveden House/Priors Way/Bray, Maidenhead, Berkshire SL6 2HP, England; in AUSTRALIA AND THE SOUTH PACIFIC by T.F.H. (Australia) Pty. Ltd., Box 149, Brookvale 2100 N.S.W., Australia; in NEW ZEALAND by Ross Haines & Son, Ltd., 18 Monmouth Street, Grey Lynn, Auckland 2, New Zealand; in SINGAPORE AND MALAYSIA by MPH Distributors (S) Pte., Ltd., 601 Sims Drive, #03/07/21, Singapore 1438; in the PHILIPPINES by Bio-Research, 5 Lippay Street, San Lorenzo Village, Makati Rizal; in SOUTH AFRICA by Multipet Pty. Ltd., 30 Turners Avenue, Durban 4001. Published by T.F.H. Publications, Inc. Manufactured in the United States of America by T.F.H. Publications, Inc.

CONTENTS

Introduction

Keeping cichlids is a fascinating pastime punctuated by moments of terrific excitement. There is no need to choose big species which are frequently destructive to plants and over-keen on digging. On the contrary, dwarf cichlids, suitable for any community tank, make ideal pets for the aquarist who is hoping for some interesting modes of behavior from his fishes. I can guarantee that keeping dwarf cichlids is never boring! With fish like the magnificent kribensis, which

Building an aquarium into a wall may make it more difficult to service—but beautiful effects can be achieved.

Some hobbyists almost build their homes around their hobby. Maintaining a huge aquarium with, for example, South American dwarf cichlids, and bringing in South American tropical plants, can make a very interesting decor. Much cheaper, by the way, than most furnishings.

can boost the beginner's confidence by becoming the first cichlids to have been successfully kept and bred by him, alongside species which can pose a problem even to an aquarist with years of experience behind him, this group of fishes has something to offer to everyone. This book is intended to familiarize the reader with the requirements of dwarf cichlids as regards keeping and breeding. Over and above that, I have endeavored as far as possible to provide such information as will enable the aquarist to sex his fishes and to differentiate the species. If he does not know his pet's name he cannot read or ask about it, and if he does not know how to tell a male dwarf cichlid from a female, he is equally likely to be in for a nasty shock. I have confined my description of the fishes to what I consider to be the vital differential characteristics, both with regard to sexes and to other species. Such dull data as fin formulae and scale counts have been deliberately excluded. The many color photographs are not just intended as an additional aid to identification but will also, I trust, help to expand the circle of dwarf cichlid fans.

The term dwarf cichlids is not a scientific one. It means a small cichlid fish, not more than about 10 cm (4 inches), which is essentially peaceful. During breeding it may be protective. Most of them take exceptional care of their eggs and young. The fish shown above is a female protecting and herding her fry.

WHAT ARE DWARF CICHLIDS?

This question is not easy to answer. The dwarf cichlids do not constitute a group that is rigidly defined either taxonomically or by their distribution. Originally the name dwarf cichlid referred solely to the colorful *Apistogramma agassizii* from South America. Later it was extended to all *Apistogramma* species. Today it is taken to denote all cichlids that only grow to a modest size—regardless of their origin or their systematic position.

Cichlids have a well-deserved reputation not only for being beautiful but also for their unusually interesting spawning behavior. On the other hand, they are also noted for their habit of changing the aquarium according to their fancy—a trait that, understandably, does not arouse great enthusiasm among aquarists. Dwarf cichlids are very different from

their bigger cousins in this respect. Almost without exception, plants are left entirely undisturbed.

Dwarf cichlids are, therefore, ideally suited for the well-planted, attractively designed community tank. As opposed to cichlids of larger size, they get on well with their fellow-lodgers. And not only that in many cases, as will be shown later, it is actually an advantage to let other fishes share their tank.

Now let us return to the question I am supposed to answer in this chapter! The term "dwarf cichlids," as employed in this book, is to be taken to mean small cichlids which on account of their behavior can be kept in well-planted community tanks without reservation. The species described here normally do not grow larger than 10 cm, and generally remain much smaller than that.

Obviously it is not possible in a book of this scope to discuss all the relevant species in detail. I have, therefore, confined myself to those which are regularly offered and which, for one reason or another, I would particularly recommend.

Your petshop can be depended upon to find a suitable aquarium and stand to fit your budget and the space available. The setup shown above features a cabinet stand, while the unit illustrated on the **facing page** is a more economical stand. Metal stands are the least expensive.

THE PROBLEM OF NOMENCLATURE

The names given to dwarf cichlids constitute an almost impenetrable jumble. For instance, it is more than misleading when (in the German language) one and the same species is referred to as "color-tailed dwarf cichlid," "wedge-tailed dwarf cichlid," "yellow dwarf cichlid," "Agassiz's dwarf cichlid," "slender dwarf cichlid," or simply as "dwarf cichlid"! All these names for *Apistogramma agassizii* have been found in popular aquarium books by me and I could "enrich" the list still

A typical coloration of a male *Apistogramma agassizii.*

further by adding the name "spade-tailed dwarf cichlid" to it. To make matters worse, the name "yellow dwarf cichlid" is applied to both *Apistogramma borellii* and *A. pleurotaenia* ! Well, one might say scientific names are used to avoid these difficulties, aren't they? And they are the same the world over. Sure. But even here we encounter a snag. For reasons I shall not go into the names are subject to frequent amendments. I have tried to use the latest names in this book.

It can be said with certainty that many specific names in current use are incorrect. Many of the *Apistogramma* species imported from South America circulate under the wrong name. Some can only be identified by the experts and a great many are still awaiting their first description. These species are, in fact, particularly difficult to describe. They often look very much alike.

ON THE BEHAVIOR OF DWARF CICHLIDS

Even just a summary of dwarf cichlid behavior in its tremendous diversity is quite beyond the scope of this book, and I want to confine my observations to those modes of behavior which belong in the reproductive sphere.

In almost all cichlids territorial behavior is a firmly established component of mating behavior. When the dwarf cichlids in our aquaria start to select and defend territories we may assume that egg-laying is imminent. In many cases the colors of these territory-defending fish have clearly grown more intense than they were before. As regards the size of a territory, the duration of its occupation, and the participation of the two sexes in territorial behavior, there exists immense variation among the different dwarf cichlids.

A male *Apistogramma bitaeniata* displaying in front of a female of the same species. In many cases the males of dwarf cichlid species are more colorful and have longer fins than the corresponding females...but not in all cases!

In the butterfly cichlid *(Microgeophagus ramirezi)* we note that both parent fishes defend as large a territory around their spawning-site as possible. In the aquarium territory size depends on several factors; the size of the fish, the degree of aggressiveness (very variable

A golden variety of *Microgeophagus ramirezi* (also known as *Apistogramma ramirezi)* has been produced by commercial breeders.

and dependent not only on individuality but on other factors as well), and on the way the available space has been divided up. The latter can also be observed in other species. When, from the fish's point of view, the interior of an aquarium has been arranged into a profusion of plants, stones, and roots the fishes are content with smallish territories. If, on the other hand, the whole width of the tank can be clearly surveyed, then territories of enormous size are established.

When the young hatch the parents defend an area around their offspring as a territory. In large aquaria one sees the adults leave the breeding territory with the young and move about the whole tank. In this way new sources of food are constantly being discovered. Frequently, however, it is not the parents who take the lead but the fry.

This is a wild *M. ramirezi* with a swarm of young surrounding it.

A head-on view of a wild-type *Microgeophagus ramirezi.* The electric blue stripe normally runs from under the eye across the upper lip and around the head. This marking disappears when the fish is not caring for its fry.

The adult animals follow the shoal and defend as large an area around the children as possible. These conditions, which can only be observed in spacious aquaria and which apply to other cichlids as well, undoubtedly come closest to the behavior of cichlids in their natural environment. A territory, then, is the area surrounding the spawning site, eggs, larvae, and fry which is defended as private. In the case of fry a territory need not be a fixed locality—it can move around.

Matters are different where mouthbrooders of the genus *Pseudocrenilabrus* (formerly *Hemihaplochromis*) are concerned. The males establish short-term territories solely for spawning. After spawning the female leaves the site with the eggs in her mouth. Not until the young, which have hatched inside her

A pair of dwarf Egyptian mouthbrooders, *Pseudocrenilabrus multicolor.*

mouth, want to leave the maternal mouth for the first time, does the adult fish select a quiet spot, there to release the offspring. The fry must

A pair of dwarf Egyptian mouthbrooders, *Pseudocrenilabrus multicolor*. The female, the lower fish, has a mouthful of either eggs or young as indicated by the distended head and jaws.

have a terrific memory for places. Whenever there is danger they are expected to return to the locality where they have originally been liberated. There, the mother stands waiting for them, ready to take them back into her protective mouth. These species, therefore, have one territory for the father and one for the mother with the young.

A female *Pseudocrenilabrus philander* watching her brood of swimming fry mostly hiding in a clump of fine-leaved plants.

Territorial conditions are equally complex with regard to *Apistogramma borellii* and a few other *Apistogramma* species. The bigger male sets up a larger territory inside of which one or more females then establish territories of their own. The females defend their territories against each other, but also against the male—except when he happens to spawn with them. Thus it is the mother's concern to look after eggs and fry. *Apistogramma* may take up so much space that there is no room left for the "master's" territory. Where this happens the females, once they have laid their eggs, frequently attack and kill their males (who are often twice their size).

It can be seen that territorial behavior is closely linked to social and brood-caring behavior. Later, I will show the dwarf cichlid fancier how the species can be grouped in accordance with their behavior. The fishes could be arranged according to the degree of their relatedness as is done by the taxonomists, but there are alternative principles of grouping that can be applied. Dividing the species into groups according to their reproductive behavior can also be a useful way of finding out more about them.

This *Apistogramma* is probably *A. ortmanni* or at least a close relative of that species.

From time to time one sees a few youngsters leave their own mother and change over to a female in the neighborhood, herself brood-caring. They are adopted without much ado.
In smaller breeding tanks females of the genus

The division into substrate breeders and mouthbrooders already goes back a long time. The mouthbrooders among the dwarf cichlids comprise *Pseudocrenilabrus* species and the goby cichlids. All other dwarf cichlids attach the eggs to a substrate and there look after them until they hatch. One could here, of course, also employ the term substrate spawners, but that would apply to the mouthbrooders as well. They, too, without exception, at least where the dwarf cichlids are concerned, lay their eggs on a substrate, i.e. on the bottom of a sand- or gravel-pit or on a stone.

A better method of grouping is the division into open and

A male Ortmann's dwarf cichlid, *Apistogramma ortmanni,* guarding eggs attached to the wall of a clay flower pot placed in the breeding tank.

Being secretive breeders, a dwarf cichlid like this Borelli's dwarf cichlid, *Apistogramma borellii,* will spawn in the least exposed area of the aquarium.

secret breeders. The open breeders comprise the fishes that, generally speaking, deposit their eggs more or less out in the open and largely without protection from view on stones, roots, or leaves. The secret breeders, on the other hand, spawn in caves or, as mouthbrooders, pick up the eggs immediately after they have laid them.

At a superficial glance this

division would seem far fetched, but it does show some interesting relationships. One striking difference, for example, is that the open breeders produce a particularly large number of small, camouflaged eggs which are guarded by both parents, whereas the secret breeders lay considerably fewer eggs which on the whole tend to be larger and more conspicuous and in many cases are looked after only by one parent.

Cave breeders are frequently very slim in build and the broodcaring parent fish in particular, in these cases invariably the mother, is often quite small. The parent is able to slip inside the cave, attend to the young, and if danger threatens to hide there as well. Conversely, the father, who gets involved in fights with rivals, as well as having to defend the territory against predators, needs to be as big as possible. This is the reason for the existing sexual dimorphism, the difference in appearance of the two sexes.

Why are the eggs of the open breeders so markedly different in number, size and color from those of the secret

A color variety of the Congo dwarf cichlid, *Nanochromis nudiceps,* that was earlier described as new species and named *N. parilus.* Note the size and location of the eggs this female is guarding.

breeders? This has got nothing to do with family relationships. The level of egg production indicates the degree of risk that exists for the brood. To ensure stocks are kept at the same strength only two, on an average, of a breeding pair's offspring will themselves have to reach the parent stage. All others perish before they get the chance to breed. High egg production, therefore, is a necessary evil for the open breeders but unnecessary where the secret breeders are concerned. The young of the latter are far less vulnerable. Likewise, the production of translucent or camouflaged egg yolks is an energy-consuming luxury the secret breeders can manage without. Their eggs can enjoy a much longer maturation period—necessary where the numbers are small to ensure that all the required nutrients are supplied. This also means the eggs will grow larger. The latter is another advantage, since the young secret breeders which are swimming free for the first time are considerably larger and hence tougher than their open breeder counterparts.

What information can the aquarist derive from this? By looking at his fish he is able to tell whether they are open or secret breeders. Fishes with a high back which are difficult to sex are almost always open breeders. Slender fish which, in addition, generally show a marked dimorphism, are cave breeders.

A close-up view of the appearance of healthy eggs of the golden dwarf acara, *Nannacara anomala.* Viable eggs are translucent, not opaque or fungused.

The culture of open breeders yields a lot of fry while spawning the secret breeders is less productive. On the other hand, one might expect that the latter are easier to breed. This is, however, not always the case—because the eggs take longer to develop, they are at a greater risk of becoming fungused.

In conclusion, I would like to

say a few words about the different family types that occur among dwarf cichlids. Open breeders form two-parent families, since the great vulnerability of the spawn makes it essential that both

The Pucallpa dwarf cichlid, *Apistogrammoides pucallpaensis,* comes from Peru. This monotypic genus is distinguished by the presence of 6 anal spines (see anal fin of a male, below). The equally distinctly patterned female is shown above.

Generally, the color pattern of a female cichlid appears most intense during the brooding period as seen in this female golden-eyed dwarf cichlid, *Nannacara anomala.* In contrast, a male *N. anomala* appears brightest during courtship and spawning, instead.

parents give their brood their constant attention. Two-parent families can be seen to be of two distinct varieties: 1) the two-parent family in the strict sense, where both parents bear an exactly equal share of

the workload (and look almost exactly alike as well!), and 2) the father/mother family where there has already been a slight shift from the center, leaving one parent with more work to do than the other, although the two partners still take turns looking after the brood, changing over at regular intervals. While the observer can already discern a degree of sexual dimorphism, the fry are unlikely to notice it. This is important. The young, instinctively, recognize their parents by specific external characteristics. It is, therefore, of undoubted advantage to them when these characteristics are exactly the same for mother and father.

A male *Nannacara anomala* as it appears when ready to breed. Note the rather drab appearance of the female fish at the bottom of the photo. Even at the same ages, the female is conspicuously smaller, too.

This male *Apistogramma agassizii*, judging from his proximity to the nest, is ready to drive away any intruder.

Where it is left exclusively to the mother to look after the brood we speak of a mother-family. A mother-family in the strict sense, however, will only be found among mouthbrooders. A great many dwarf cichlids form a sub-type of this mother-family which can be described as a male/mother family. As before, the broodcare is left exclusively to the mother, but by defending his territory the father is indirectly helping to secure the survival of the young.

These divisions into family types are highly schematic. It is, therefore, not only because of the possible transitional or intermediate types that they have to be employed with care. To give just one example, in tanks with sufficient space *Apistogramma agassizii* usually forms a typical male/mother family. Sometimes, however, the male does not take the slightest interest in his territory, i.e. we have a mother-family in the strict sense! Then again, Pinter has repeatedly observed modes of behavior where this species is concerned, which can only be interpreted as coming from a father-mother family. And recently Dr. Frank, Prague, wrote to me that he knows of many cases regarding *A. agassizii* where the male alone looked after the brood—a father family!

The requirements of the individual species and genera vary so much that only very general advice can be given here. More detailed information will be found in the descriptions of genera and species.

Most aquarists will want to accomodate their dwarf cichlids in a community tank. They can be kept with the great majority of aquarium fishes without reservation. It is important, however, to meet the requirements of the cave breeders and provide possible hiding-places. That many species make special demands on the properties of the water should also be borne in mind.

I would like here to suggest how to set up a South American tank with dwarf cichlids foremost in mind. Before we decide what to put in the tank we should ask ourselves whether we are able to provide our fish with water of the correct quality. The water should be fairly soft—I would regard 14 degrees of German hardness as the extreme maximum and a hardness of less than 10 German degrees as ideal. Where the water comes out of the tap too hard these values can only be reached by means

A tank with a small surface area is not suited for most fishes, except perhaps the very small species and air breathers like the anabantoids.

An odd-shaped tall aquarium may be appropriate for helping to create special decorative effects, but not for breeding most fishes, including cichlids.

Regardless of the shape of the tank selected, it should be provided with good filtration, heating and lighting equipment and should have its water quality checked on a regular basis.

of a softening filter or by adding distilled water. The use of rain water has to be considered very carefully in most areas—all too often it is already heavily polluted when it falls to the ground.

As most tap water is slightly alkaline, a degree of acidification with peat extract or other media offered by aquarium stores for the purpose is recommended. The application of a filter is advisable, although not absolutely essential. An aerator should be present. Our thermostat is adjusted to 25°C. When a partial change of water becomes necessary it does not matter if the soft water we add is so cool that the tank temperature falls by 2 or 3°. In fact, this simulates a tropical cloudburst and stimulates the vitality of our fish. Often the fishes begin to spawn as a result.

The size of the tank depends on one's financial resources. The bigger the aquarium, of course, the greater the pleasure it later

Ideally, the plants selected for a dwarf cichlid tank should match those found in the fish's natural environment.

gives us. On the other hand, to supply a large tank with soft water may well turn out to be difficult and costly.

I shall refrain from making detailed suggestions as to planting. I would not consider it a serious mistake if a "South American tank" also contained Cryptocorynes from southern Asia. What does matter, however, is that plants, roots, and stones are arranged in such a way as to leave several "recesses". Even in a tank with a length of no more than

Airstones of varied shapes and sizes are available to suit a hobbyist's tank set-up and requirements. The type shown here can be placed in the background, perhaps hidden from view by a bank of plants, rocks, or other ornaments.

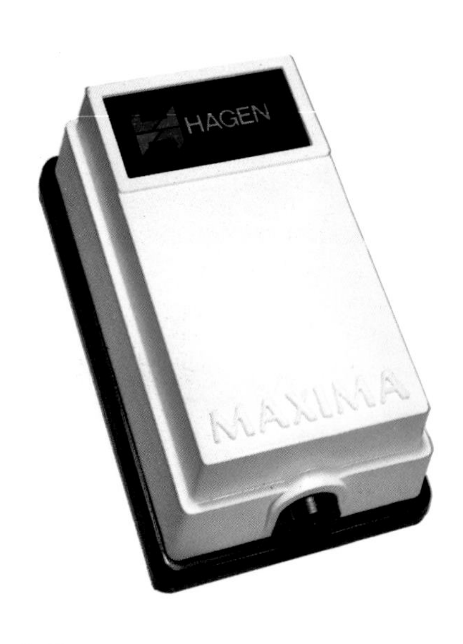

A vibrator air pump is inexpensive to purchase and operate. However, be sure to get one that is powerful enough to operate all the equipment it is intended to run.

40 cm this should be quite easy to ensure. These spaces are particularly vital near the bottom since that is where, virtually without exception, our dwarf cichlids spend most of their time.

Equally important is the provision of an adequate number of caves. Whether we use flower pots or coconut shells for the purpose is a matter of preference. Pet shops sell many ornamental caves. This type of cave does, however, have the advantage

Top photo: If you elect to use an undergravel filtration system in your aquarium, the pet shop dealer will be the best person who can show you how to install and handle such equipment. The power head used for increasing the flow rate of an underground filter is shown here. **Bottom photo:** Today, filter boxes are so designed that the component parts are easy to assemble for cleaning and replacement of filter material.

Your dwarf cichlids will not hesitate to use this store-bought tank decoration for spawning.

that it can easily be moved an inch or two should the need arise. It also makes it easy for us to take the eggs out of the community tank and transfer them to a special tank for artificial incubation.

In a particularly attractive aquarium, designed to resemble the natural environment as closely as possible, one could build

Some commercially made rockwork items may look like the real McCoy, although they are constructed of man-made materials.

caves from roots and stones instead of using flower pots and coconut shells. But—caution, please! An artificial cave made of stones can collapse! One has to realize that the fishes want to have a say in the design and size of their cave. They want to remove fine sand from the cave in order to increase the available space. Ornamental ceramic caves are best partially buried in the sand. With stone caves there is obviously a great danger that the fish might burrow under them. The stones making the caves must not lie on sand!

Let us not forget the open breeders. For them we put a few stones—from the size of a ping-pong ball to that of a child's fist—into the aquarium. It should soon become apparent which of these spawning sites the fishes prefer. On no account must stones containing calcium get into the tank! They would quickly increase the water hardness at an alarming rate. Calcium is easy enough to detect. All we have to do is to place one drop of hydrochloric acid on the stone. If the acid foams as a result, then the stone contains calcium and is unsuitable for our purpose.

Now to the inhabitants! Their number depends on the size of the aquarium. A tank with a length of 1 to 1.25 m, properly divided up, could accomodate 2, or even 3, cave-breeding *Apistogramma* species quite comfortably. It is important, however, that members of the different species can readily be distinguished. It is not always easy to tell them apart, particularly where the females are concerned. One should, therefore, avoid mixing the different species of the *borellii* type. A favorable combination would be one *Apistogramma* species from the slender group *(A. agassizii, cacatuoides)* and one from the group of the more thickset species *(A. borellii, macmasteri* types*)*. Wherever possible, the aquarist should purchase one male (only) and several females per species.

We may combine the cave-breeders with two open

Some suggested designs of preparing a tank for housing and breeding dwarf cichlids. Plants, stones, and roots are utilized in combinations to suit open breeders or cave breeders.

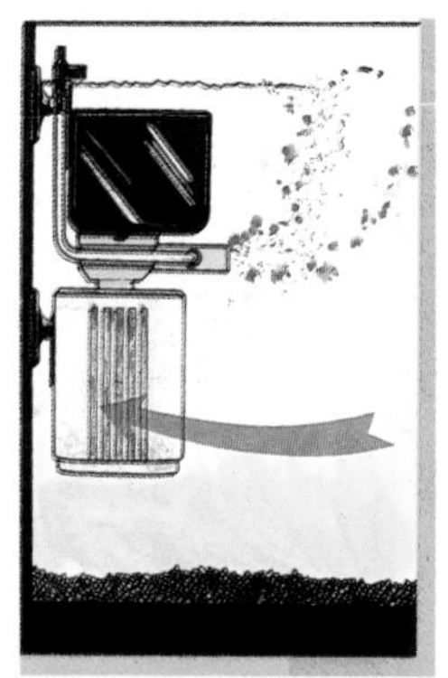

A diagram showing a submersible power filter and filter box. The arrow shows the direction of water flow.

breeders—for example pairs of *Aequidens curviceps* or *M. ramirezi.* A dwarf cichlid tank with a mixed population like this is guaranteed to be interesting and to give pleasure. I would, however, add a few other fishes as well—some splash tetras, silver hatchetfish, or other characins which have a preference for the upper water layers. They liven up the upper region of the aquarium, but that is not their most important function. Associated only with each other, dwarf cichlids are often very shy and hide at the slightest movement outside their tank. The characins calmly swimming about in the upper range give the dwarf cichlids a feeling of security according to the motto, "If they don't feel threatened, then there shouldn't be any danger for us either."

Now a tip for those who have to make do with a small tank or for aquarists who are finding it difficult to obtain soft water. It is perfectly possible to set up a 20-liter tank with a length of about 40 cm in such a way that even the biggest *Apistogramma* species can be kept and bred in it. I myself have tried it out with *A. bitaeniata,* and others. No doubt some *Apistogramma*

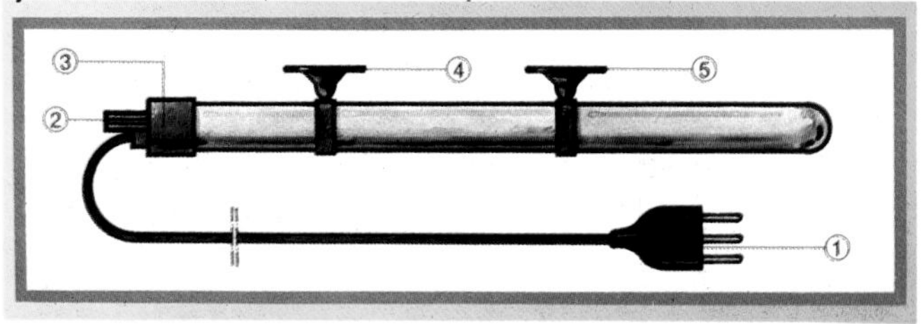

A submersible heater, being located below the water, reduces considerably the clutter of aquarium equipment normally seen behind a tank.

fanciers will be horrified to hear this, the general message being, after all, that these fishes need vast territories and therefore, at the very least, a "meter tank"!

The population of this mini-aquarium must, however, not exceed one male and one or two *Apistogramma* females, along with three to five cardinal tetras, *Cheirodon axelrodi;* which should be included for reasons mentioned above.

I fill roughly half the mini-tank with fist-sized, rounded stones, leaving relatively large gaps between them. In the two back corners the stone arrangements are almost high enough to reach the water level. In the two front corners they only come to about half the height of the tank. In the foreground and in the central area, or, for aesthetic reasons, perhaps slightly off center, I leave a small space, part of which is to accommodate the plants.

It is important that the stones are wedged in tightly so that they cannot collapse. The sand—as fine as possible and free from calcium—is added afterwards. Now there is no danger of our cichlids digging under the stones and bringing them down.

Cardinal tetras, a schooling type of fish, will not interfere with the activities of many dwarf cichlids that generally stay close to the bottom and hide behind rocks and plants.

Set up like this, even a very small tank enables the fish to keep out of each other's way. A mini-tank arranged in this way also, and most importantly, enables the females to seek refuge from their males. They will find tiny "caves" (nooks and crannies) into which the bigger males cannot follow them. Another advantage of these "stone tanks" is the comparatively small water requirement, a major part of these very small tanks being taken up by stones. This can be a great bonus when it comes to the preparation of soft water.

There are two important points to remember as regards these smaller tanks. The stone constructions must be absolutely safe from collapse, and the aquarist must be aware that a small quantity of water becomes polluted more quickly than a large one. A weekly nitrite test should, therefore, be carried out and the water may have to be changed at relatively frequent intervals. (This is important also because a filter is unlikely to be connected to so small a tank.) However, in this case we can change the water without bothering to take out decorations, heater, or fish. We simply carry out partial

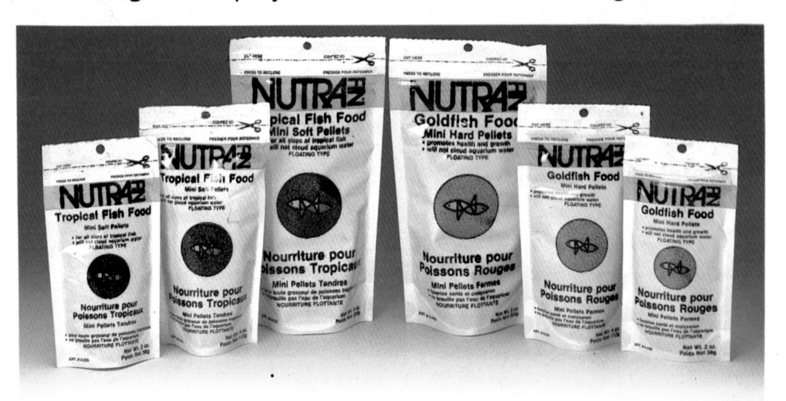

Pelleted fish food, like flake foods and other dry prepared foods, is almost an indispensable staple for hobbyists and fish breeders all over the world. Not requiring refrigeration, it can be fed to fish anywhere. **Facing page:** Tank setups possible for open breeders that do not require caves, just some stone slabs or pieces of rocks.

water changes—that is, about one third of the water is drawn off and replaced at a time.

An African tank is set up according to the same principles as the South American tank. That is: good spacing by means of stones, plants, and roots, and the provision of breeding caves and spawning stones. Unless we decide to stock the tank with *Nanochromis* species (native to the Congo), we do not have to worry about water hardness. This is one advantage over the South American tank. As cave-

breeders we can select the magnificent "King kribensis cichlid" *(Pelvicachromis pulcher)* which in their native West Africa occur along the whole course of large rivers, often as far as the brackish zone, and therefore also tolerate harder water very well.

The open breeder I would recommend here—as the counterpart of the butterfly cichlid, to which it actually bears a close resemblance—is *Anomalochromis thomasi.* Another fish we could add is the dwarf Egyptian mouthbrooder

Facing page and above: Processed fish foods, regardless of the form—flakes, pellets or tablets, freeze-dried or frozen—usually have similar nutritive value. However, the stage of development and eating habits must be considered. Fry have food requirements different from adults, surface feeders will utilize floating food best, tablets may take a while to dissolve, etc.

Pseudocrenilabrus multicolor. This species will ensure that there is life and variety in the tank. These fish are normally kept in water of medium hardness. The tank could further be stocked with Celebes rainbowfish. Though not Africans themselves, these fish get on well with the African dwarf cichlids mentioned above and prefer medium to hard water. They are lively fish of the upper water layers.

From time to time an observant keeper will notice that the cichlids in his community tank are already busy looking after eggs, larvae, or even free-swimming fry. But only in sparsely populated aquaria is there any chance that at least a few fry might survive the harassment by the other occupants of the tank. In such cases it will generally be necessary to provide extra food in the form of brine-shrimp larvae *(Artemia salina)*, which are blown towards the shoal of fry by means of a tube. It is better to transfer the brood to a special tank.

To transfer the eggs to a raising tank, a small, deep bowl is placed in the community tank, the spawning

A strong flow of oxygen-rich water can save a clutch of eggs from fungusing. Be sure that the eggs are not directly agitated, however.

Your local pet shop carries medications for whatever ails your fish—fungus, parasites, malnutrition, etc.

stone or cave with the eggs is put into the bowl, and bowl and contents are lifted out of the aquarium. By the same method, the stone is then put into the breeding tank. Most of the water in this tank should come from the community tank. Ideally, this water should now gradually be changed for new water. The outlet of the aerator is placed as close to the eggs as possible so that the latter are receiving a constant supply of new water. The eggs should be hanging down at a slight angle, so that

the brood can sink to the bottom immediately after hatching.

To prevent, in as far as possible, the eggs from becoming fungused, we pour a little fungicide into the water. Some breeders prefer to add methylene blue (till the water shows a slight blue turbidity). An effort should be made to remove any fungused eggs (which can be recognized by the white discoloration) with a fine needle or forceps, else healthy eggs will become infected. Unfortunately, this tends to be a laborious, tiring task.

Sometimes there are complications when the young hatch. This varies from one species to another. The parents of *Apistogramma* species in particular often help their progeny into the world by sucking at the egg membranes until they burst. We are not able to imitate this. It may be better, therefore, to leave the

A female *Nannacara anomala* hovering above her eggs (upper photo). The same female "peeling off" the membrane from the last egg with embryo (lower photo).

Mr. Karl Schnell, managing director of T.F.H. Publications (in Australia), adjusting the air pump so as to cause a stream of water to flow in front of eggs (in this case, angelfish spawn on the surface of a leaf affixed to the side of a glass jar). With some ingenuity it is possible to hatch cichlid eggs without parental care.

brood in the care of the parents until they have hatched.

Once the babies have emerged the mother gathers them together in small depressions. It is then easy to siphon them off and transfer them to a separate tank. Any fry already swimming free are best caught by the following method. Aided by the mother or both parents, the young assemble in the spawning cave or some other place, there to spend the night. If we switch the aquarium light back on again late at night and go to work immediately with our piece of tubing, we can surprise the young when they are still half asleep and transfer them to the breeding tank without much difficulty.

Results are often better if the breeders are spawned in smaller tanks of about 50 liter capacity. Where a mother family has been formed (and by this I also mean a male/mother family) the father animal is netted out after spawning. Wherever possible the mother should remain with her brood even when the fry are already free-swimming. A good rearing-food to offer in all cases are newly-hatched brine shrimp *(Artemia salina)*. The nauplii of *Artemia* are a popular first food with all dwarf cichlids and promote growth. The brine shrimp are simple to breed—with a suitable bottle and an aerator easy success is guaranteed.Pet shops sell brine shrimp hatcheries.

This *Nannacara anomala* female is keeping her eyes on her "hopping fry" that are not yet ready to swim freely.

A lot of mistakes are made when it comes to rearing the

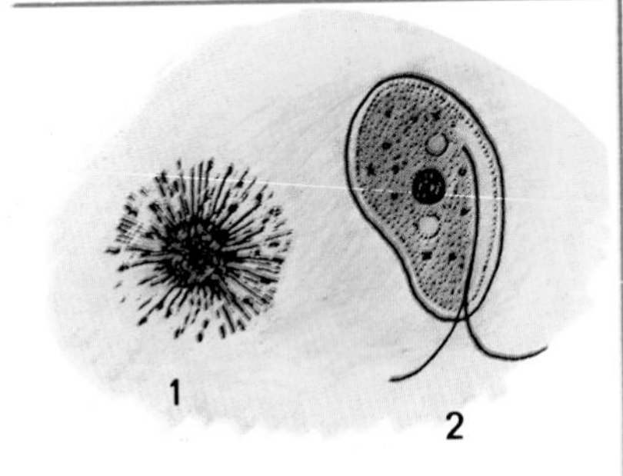

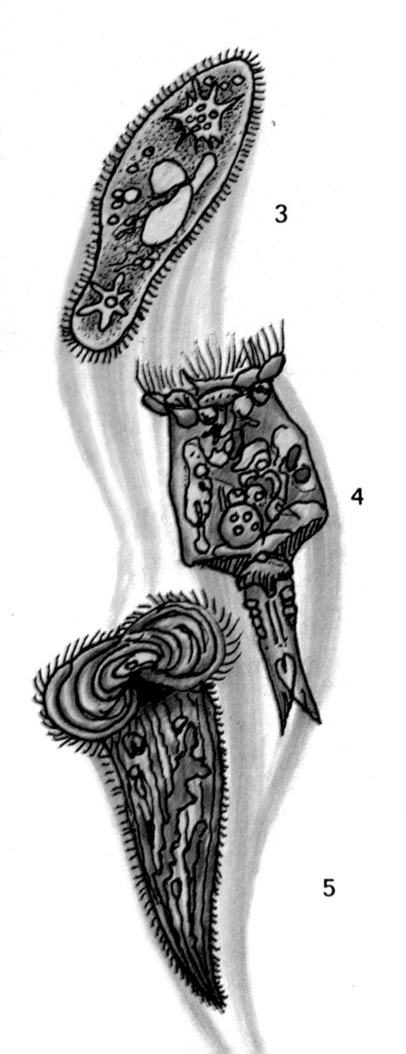

fry. The young need adequate quantities of very small food animals and regular changes of water. Otherwise they are underweight and sickly and in no fit state to be offered to anyone. There can be no doubt that it is well worthwhile to do the job properly. Unlike the tank-raised specimens of large cichlids, dwarf cichlids are never difficult to sell. Dwarf cichlids are popular!

A word of warning to those who want to leave the fry in the community tank with their parents for a bit longer. Be careful when putting *Daphnia* into the tank! Admittedly there is no danger at this stage that the cichlid mothers with their excessively powerful maternal instinct will mistake the water fleas for young cichlids and adopt them as such, but the fry themselves frequently get so puzzled by the water fleas jumping about chaotically in the tank that they latch onto them and, without taking any more notice of the mother, spread themselves out in the

Some infusorians that can be fed to newly hatched fish fry: 1. Radiolarian, 2. Freshwater flagellate, 3. Paramecium, 4. Rotifer, and 5. Vorticella.

aquarium. They then only too easily fall prey to the other occupants of the tank.

In conclusion I would like to point out a few species which are suitable for novice breeders. I am well aware that the beginner sometimes succeeds in breeding the most difficult species at the very first attempt, and that an "old hand" occasionally comes to grief over the "easiest" fishes. There is many a case in point I could quote here! However, the following species are relatively easy to breed: *Nannacara anomala*, *Aequidens curviceps*, *Pelvicachromis pulcher*, and the *Pseudocrenilabrus* and *Julidochromis* species. The *Apistogramma* species are a little less easy, but with *M. ramirezi*, *A. borellii*, and *A. cacatuoides* even the novice may well achieve good results. Don't be afraid of the other species, though! With a little patience and hard work it is perfectly possible to propagate all the species mentioned here.

The genus *APISTOGRAMMA*

By 1988 the number of *Apistogramma* species had risen from five (1906) to over

On account of her dark markings and special movements, a female *Nannacara anomala* is able to keep her swimming fry from straying far.

60, and still more species will undoubtedly be described in the near future.

I decided against including a list of the species described so far. It is doubtful that it would have removed any of the uncertainty attached to the determination of *Apistogramma* species, and, what is worse, in some case it might have led to a certain deception. The reader must be aware that a considerable number of forms available today still await scientific description. In spite of this, many of these fishes are sold under (false) scientific names.

Dr. Froehlich returned from a journey to Colombia with several species he had caught which had not yet been described. Every one of them was typical for a very specific and very limited range. These

Finding *Apistogramma* specimens in the wild is not too difficult, but finding their scientific names can take years. The unnamed *Apistogramma* shown on this and the facing page were collected by Dr. Herbert R. Axelrod in South America. (1, 2) collected in 1975 in Rio Negro, near Manaus; (3) was from the Rio Tefe, in 1974; (4) was taken from Humaita, Brazil in 1978, (5) collected sometime in the 1960s (possibly *bitaeniata)*.

1↑ 2↓

3↑

4↑ 5↓

localities were only about 20 to 50 kilometers apart. What profusion of new species may yet be expected to reach us from South America?!

The determination of round-tailed species in particular *(pertensis, commbrae, pleurotaenia,* and many more *)* poses great problems. Anyone wishing to keep and breed these species must be content with the general information given below. But there also exists highly characteristic and fairly widely distributed *Apistogramma* species about whose classification nobody has the slightest doubt. The most important of these I have dealt with individually. Particularly bizarre to look at are the high-finned forms. The dorsal fins of *A. bitaeniata* and a few other species can be of the same height as the body, or even taller. It is not the fin rays which give this effect, however, but the greatly extended membrane between the spines. In other words, the tips of the dorsal fins are soft in these species!

Apistogramma males usually occupy fairly large territories inside which several females

The upper fish is a male *Apistogramma pertensis*, the lower one possibly *A. hippolytae.*

This fish is still waiting for a name (new or old) from experts on dwarf cichlids.

establish smaller breeding territories. This behavior precludes any real pair-formation. The extent to which this applies, however, varies from species to species. While, for example, in *A. trifasciata* pair formation is almost completely absent, it is the rule in *A. commbrae.* Division of duties seems to mean a less rigid pair-formation and, at the same time, an increase in sexual dimorphism. Conversely, where the sexes resemble each other as closely as they do in *M. ramirezi* they enjoy almost complete equality with regard to territory and brood-care. I would like to add, however, that the special behavior of these species can also show considerable variation between individuals as well as being strongly influenced by external conditions.

A few hints on breeding: The decisive factor in spawning is the female's readiness. A female wanting to spawn will do so even under relatively poor conditions. Sometimes even the side of a bare bowl is considered a suitable substrate. Fishes which after a prolonged period of hardship are suddenly able to enjoy good conditions again often turn into particularly keen spawners. This is likely to be one reason why imported animals generally spawn much more readily than do tank-bred fish.

Ideal spawning places have been found in flower-pots, either halved or, if left whole, placed horizontally in the sandy bottom. The females like to further improve on the size of their caves by digging. Other breeders prefer flowerpots from which the base has been removed, putting these into the tank upside down. Many fishes show a preference for the latter arrangement, as it means they are hidden from view. But this can have obvious disadvantages for the observer—and often for the breeder as well.

For the great majority of species it is absolutely essential to have soft breeding water. But even then the eggs are all too prone to fungus. To prevent fungus infection use one of the commercial fungicides, or simply methylene blue and make sure there is a strong water current. Very many *Apistogramma* species originate from South American brooks with a relatively strong current!

A tank setup for breeding dwarf cichlids. It has a protected site for the nest (a clay pot lying on its side); a filter (sponge filter) that will not suck in small fry; and a place where a female can seek protection, a floating piece of styrofoam.

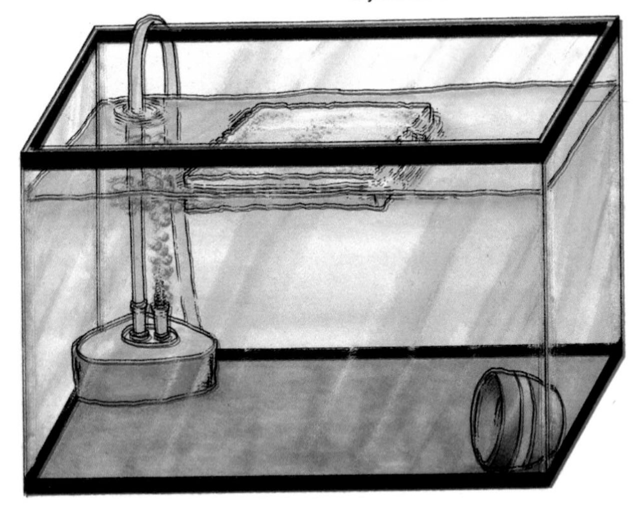

Apistogramma gibbiceps (male) photographed by Hans-Joachim Richter in Leipzig, East Germany.

Chemicals for the prevention of fungus in eggs—and indeed any chemotherapeutic agents—must be applied with care, however, as our fishes are very sensitive to them. Treatment should commence with only one third the recommended dosage!

Many mothers devour their own eggs for no apparent reason. When the light is switched off at night the eggs are still there and when one comes to have a look at them the following morning they have disappeared. One breeder advised me, recently, that egg-eating could be prevented by leaving the light on continuously. He maintained that this was almost always effective. I myself have not tried out this particular method yet but would like to take this opportunity to pass on the advice.

Apistogramma steindachneri (male).

APISTOGRAMMA STEINDACHNERI
Steindachner's Dwarf Cichlid

Characteristic of this species which, by *Apistogramma* standards, is endowed with a particularly long snout is the markedly extended (above and below) tail-fin in older males. This applies to other *Apistogramma* males, too, but the latter possess an additional differential characteristic, visibly extended membranes between the dorsal fin rays which are absent in *A. steindachneri* and the very closely related forms *A. wickleri* and *A. ornatipinnis. Apistogramma wickleri* is presumably only a very beautiful race or aquarium variety of *A. steindachneri.* The horizontal stripe on the body terminates in both forms as a spot at the caudal base.

A. steindachneri males can reach a length of up to ten centimeters. The females, with a maximum length of 6 to 7 cm, also attain a considerable size. In spite of this they are comparatively peaceful fish and plants are safe with them too. The males have shiny scales of a blue to greenish color.

The melanin pattern, i.e. the black and dark-brown spots and stripes, is dependent on

Upper photo: *A. steindachneri*, possibly a young male (note slightly pointed dorsal fin). **Lower photo:** *A. steindachneri* (female).

mood and can, therefore, vary a lot in this as in all other *Apistogramma* species. Hence, when describing the other species, I shall point out only those color patterns that are especially characteristic. Here, however, are a few examples of these markings

Fishes which are not feeling

well or have in some way been oppressed show, apart from the stripe on the cheeks which is is almost always present, their horizontal band and, in addition, six vertical stripes on the body. Territory-owning males, as also do the brood-

Apistogramma steindachneri (female). The markings are not permanent; the spots (upper photo), cross bars, and longitudinal band can appear and disappear, with changes in mood. In another female (lower photo) the cross bars are very noticeable, a lateral spot not evident.

A mature male *A. steindachneri.* Males can grow to a length of 10 cm. Note the pointed ventral and dorsal prolongations on the tail fin; females have rounded tail fins.

caring females, show a single central spot on the body and another spot at the caudal base. The longitudinal band disappears completely. During courtship and spawning both the distal and central spot may become invisible. This description is meant to show the hobbyist, who does not usually occupy himself with preserved specimens, that he cannot rely on melanin markings all that much when determining this *Apistogramma* species. Much more important differential characteristics are body and fin development.

The fish are fairly peaceful, but on account of their body length need a sizeable territory nonetheless. Breeding the species is relatively easy. The fishes prefer to lay their eggs in caves, but my own fish have been known to spawn under a *Cryptocoryne* leaf, too. The females, at this stage yellow in color with beautifully contrasting black markings, defend their brood fiercely. They attack their enemies with sudden abrupt movements, the head held low in a characteristic posture.

Sometimes the male takes a very active part in defending the territory. The female acknowledges the male as an equal partner in brood-defense, particularly when the enemy exerts a lot of pressure. I even saw one pair where individual territorial duties were the exact reverse of those normally described with regard to *Apistogramma* species. The male stayed with the eggs while the female prevented "the enemy" from crossing the pair's territorial boundaries. Of course this behavior is not typical for the species, but it does show that anything can happen where *Apistogramma* species are concerned and that it would be a mistake to generalize too soon.

APISTOGRAMMA TRIFASCIATA
Three-Stripe Dwarf Cichlid

The males of this small species attain a total length of 6 cm maximum. In the aquarium, however, a length of 5 cm would already be quite considerable. The females remain a centimeter shorter.

In males as well as females the caudal fin is rounded at the end. In old males it may sometimes become angular both above and below, though without developing obvious tips. A special adornment of the male is the prominent dorsal fin which looks like like the head-dress of an Indian—the tips of the membranes between the fin rays are of a magnificent orange or fiery red and very long.

The species owes its scientific name to the three stripes on the body. The first stripe starts on the tip of the snout, goes through the eye, and peters out at the caudal base. The second stripe is the cheek stripe characteristic of the *Apistogramma* species, which originates at the lower margin of the eye and slants to the back, across the gill cover. Roughly parallel with the cheek stripe, sometimes barely visible, runs the third band. In the form of a narrow stripe it connects the base of the pectoral fin with that of the anal fin. There are supposed

Facing page: *Apistogramma trifasciata*, (1) male (2) female (3) male and female. This species is widely distributed and three subspecies have been described.

1

2

3

to be three subspecies of *A. trifasciata.* From the subspecies *haraldschultzi* we can readily distinguish the original form by the origin of the body stripe. In *A. trifasciata haraldschultzi* the "*trifasciata*-band" starts slightly above the insertion of the pectoral fin. A better differential characteristic, however, is the big horizontal band which in the original form starts at the upper lip while in the subspecies it clearly has its origin above the upper lip. The original form, which strictly speaking should be called *Apistogramma trifasciata trifasciata,* is found in the upper range of the Rio Paraguay. *A. t. haraldschultzi,* on the other hand, originates from beyond the watershed, from the Amazon range—i.e. the region of the Rio Guapore, one of the rivers dividing Brazil and Bolivia. East of this area,

The characteristic three bands are evident in this mature male *A. trifasciata.*

The very dark middle band seen in this female *A. trifasciata* guarding eggs will be reduced into a round spot at the middle of the body later when the fry are hatched and free-swimming.

that is in the central Amazon region, is the home of another subspecies, *A. trifasciata maciliense.* Some authorities regard the three races as independent species, and there are sound enough reasons for this way of thinking too. It is really just a matter of personal preference which scientific view one decides to adopt.

Quite a lot is known about the native water of *A. trifasciata haraldschultzi,* though regrettably this information is exceptional where *Apistogramma* species are concerned. Hermann Meinken, by whom the species was first described, states that as early as 7 a.m. the water temperatures in this particular spot already lay within the range of 23 to 25° C. He goes on to say, "Due to the scorching sun the water temperature rises considerably during the day," and stresses, "that our *Apistogramma* is not at home in the darkness under the cushions of floating grass but among the tangled mass of plants in the sun-scorched lagoons."

In accordance with this report, the species should not be kept at too low a water temperature—26 to 30°C should be offered to these fish. Equally important for *A. trifasciata* is the provision of soft, slightly acid water, and water changes should be carried out with special care. To me these fish seem even more sensitive to their environment than the other *Apistogramma* species. However, once the water requirements have been understood and they are given plenty of live food, *A. trifasciata* is easy to breed. The characteristic social stucture seen in the genus *Apistogramma,* i.e. one male "super-territory" and several sub-territories for the females, is particularly apparent in this species. These dwarf cichlids, too, can of course be kept inside those miniature rock aquaria described earlier, but if one wants to study the typical family structure of this species

The full color and body markings are all displayed in this male *A. trifasciata.*

An alert female *A. trifasciata* (on the right) watches every movement of the male during the breeding cycle.

the tank should be as big as possible. Then one will also be able to observe how the territorial females, who are anything but on friendly terms with each other, kidnap one another's young. This even applies to females who have never spawned. On first thought this may seem strange, but it make sense when one knows that the broodcaring instinct in the females is so strong that they sometimes even look after water fleas and *Tubifex!* The layman will suspect they are merely defending their food with particular jealousy, but the experienced aquarist can see that these animals are suddenly wearing their brood-caring dress! This behavior, incidentally, is shown by all *Apistogramma* species.

Females whose brood is already free swimming move about in the tank and may migrate from one male's territory into that of his

neighbor. Males, too, may display a strong brood-caring urge in which case they often take it upon themselves to look after the older fry after ousting the mother.

These photos illustrate the changeable aspects of the markings and coloration that mature and live males of *A. trifasciata* can show. Even when the same fish is photographed in rapid succession, it will be almost impossible to get the same appearance.

At a water temperature of about 27°C the young start to have fights with each other after a maximum of 15 days. A day or two later it is no longer possible for the mother animal to keep them together.

An adult female *Apistogramma trifasciata* guarding her pink eggs attached to the inner wall of a coconut shell.

APISTOGRAMMA BITAENIATA—Banded Dwarf Cichlid

The banded dwarf cichlid is quite similar to the species just described but larger and more splendid. The males reach a total length of up to 9 cm—considerable for a member of the genus *Apistogramma!* —and a body length of 6 cm. The females, with a total length of 5 cm, remain much smaller.

The males' most signficant adornment is the gorgeously colored dorsal fin with its greatly extended membranes between the rays, orange to red at the tips. Along with the forms with bluish-grey fins

Besides the normal form with blue-green fins, a form of *Apistogramma bitaeniata* with bright red tail, dorsal and anal fins exists.

Apistogramma bitaeniata (normal form), male.

there have recently also been appearing fish with bright-red caudal, dorsal, and anal fins. They are easily confused with *A. sweglesi.* Considered as distinct from *A. bitaeniata,* however, *A. sweglesi* has 16 spiny rays in the dorsal fin—the "banded dwarf cichlid" having 15. A "red *bitaeniata*" I closely examined, however, turned out, apart from differing in a few other respects as well, to have only 14 spiny rays, a very unusual number for an *Apistogramma* species. This would speak for a separation of this form from *A. bitaeniata.* As an argument against it, however, we have the reports of breeders who maintain they have raised individual broods where blue forms occurred side by side with red ones.

A. bitaeniata is distinguished by having three to four ruby-colored cross-bands at the end of the central rays of the caudal fin. These are absent in females and younger males, however, and only hinted at in the (?) "red *bitaeniata*." Characteristic of *A. bitaeniata* males is a band composed of partially brownish-red scales, which starts at the posterior edge of the eye and, almost going as far as the back, runs along the caudal base. In the (?) "red *bitaeniata*" this band is (of course!) not as obvious.

A particularly prominent characteristic of the species is several bluish-gray stripes in the anal fin which run in the direction of the fin rays. These markings are already shown by younger males. Occasionally a hint of them may also appear in a female. On the whole, however, it is as difficult, as with the great majority of *Apistogramma species,* to pin down characteristics that can be said to differentiate the females of *A. bitaeniata* from females of any other species of this genus—apart from fin ray and scale counts, of course. Older females are, nevertheless, often quite easy to differentiate from the round-tailed species. In an older female the caudal fin ceases to be rounded, looking as though it had been cut straight. In some specimens the tail-fin looks almost trilobate.

The species is said to originate from the central Amazon. It makes a few demands on water quality

A. bitaeniata, female, has deposited the eggs within the cavity of a big stone.

A. bitaeniata male showing his most brilliant breeding colors while resting above a clutch of eggs.

which have to be met if one wishes to keep it for a longer period or even intends to spawn it.

Hardness and pH are not the only decisive factors as regards water quality. Rather, a combination of factors are important which still need to be determined precisely—not only for *A. bitaeniata* but for the other *Apistogramma* species as well. Meanwhile I would remind the aquarist of the dangers of drugs and water pollution—and perhaps also of the potential danger of *Tubifex* and red mosquito larvae which frequently come from contaminated waters and can store toxic substances in their bodies!

For *A. bitaeniata* I would recommend a water hardness of 8 German degrees or less and a neutral or very slightly acid pH. The thermostat should be set at about 28° C.

A. bitaeniata is more difficult to breed than many other *Apistogramma* species. The female sticks her eggs to the ceiling or vertical wall of a cave. The number of eggs usually remains well below 100. If the spawning cave is too small for the large male, he fertilizes the eggs by discharging the sperm just outside the cave and then sending the spermatoza into the cave with powerful beats of his tail fin. This can be observed in other *Apistogramma* species, too, as can also the following phenomenon. The color of the eggs can vary tremendously depending as it does on the food the female has eaten before spawning. After a *Cyclops* diet the eggs will be bright-red. After certain varieties of dried food they become whitish-gray or, quite often, yellowish-brown.

The fry hatch after about three days, and are free swimming after another five days. From then on, if given newly-hatched nauplii of *Artemia,* they are easy to raise. At the age of six months they are already able to reproduce themselves—provided they have been properly kept and cared for.

APISTOGRAMMA SWEGLESI—Swegles's Dwarf Cichlid

This dwarf cichlid was only seldom imported. The species is frequently confused with the red strain of its close relative, *Apistogramma bitaeniata.*

The fish come from the area around Letitia on the upper Amazon. There they live in the open spots of rivers with a dense surface- and underwater vegetation.

The slender males are reddish-brown on the neck and back and brownish to olive-yellow in color towards the flanks. The underside is paler still. Their fins are gray and show a dark-brown border in part. In contrast with nearly all the other related species, the lower part of the cheek band is clearly bent forward.

Apistogramma sweglesi males attain a length of 7 to 8

Apistogramma sp. affin. *taeniatum,* male above, female below.

cm, of which 1/5 is attributable to the extended, doubly pointed tail fin. The females, with a total length of 5 cm, remain markedly smaller.

Upper photo: *Apistogramma gibbiceps* has fourteen dorsal spines, a distinct longitudinal stripe, oblique markings on the belly, and a double-pointed tail. **Lower photo:** *Apistogramma pertensis,* male.

APISTOGRAMMA PERTENSIS
Amazon Dwarf Cichlid

For a long time identification of *A. pertensis* was very difficult. The lower photo on p. 64 shows the real *pertensis*: slender body, with a very high dorsal and stripes in the caudal fin! More about this fish later.

Kullander is at present working on a revision of the *Apistogramma* species. According to him *A. pertensis* has never been among imported aquarium fishes, at least not up to 1973. Various fishes until now referred to as *"A. pertense"* in aquarium stores, magazines, and books belong to other round-tailed species, most of which are in fact still awaiting their scientific description! No wonder, then, that reports on keeping and breeding are so conflicting!

Recently Froehlich was able to bring back fishes from the clear-water region of eastern Colombia which obviously belonged to different species,

This fish was collected from Colombia and was identified as *Apistogramma pertensis*. It is probably *A. macmasteri* or a species closely related to it.

yet counts and measurements in every case were the same as those found in the original description of *A. pertensis.* So now the picture is completely confused! Not only differences in appearance were discovered, but also in behavior. All the fishes originated from tributaries of the Rio Meta, itself a tributary of the Orinoco.

Facing page: A typical Apistogramma biotope in Colombia, about 100 km east of Villavicencio near Puerto Lopez, in January. The shallow water of the inlet, only, about 10-30 cm deep, is stirred by collectors and appears a brownish. Here many *Microgeophagus ramirezi* live and rear young. Temperature of the water in the inlet 30-35° C. Photo by Dr. F. Froehlich.

This diagram shows a typical cross-section of a creek in Colombia and depicts the distribution of the species. Species of *Apistogramma* are found preferring the edge of the creek under the overhanging grass sod andd roots. Drawing by S. Haag after an original by Dr. F. Froehlich.

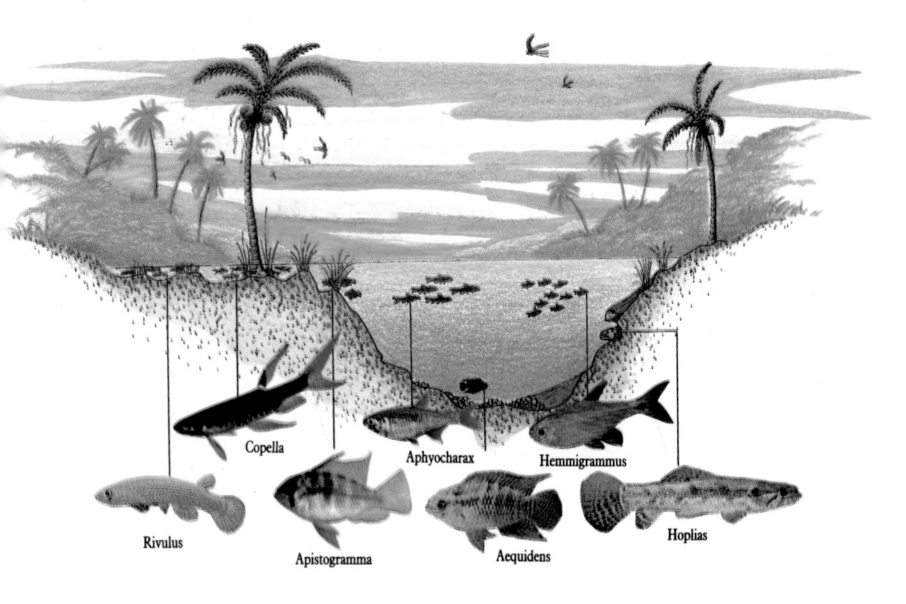

The information provided by Froehlich on the distribution of his *Apistogramma* species is so interesting that I would not want to withhold it from the reader. The accompanying illustration shows a cross-section through one of the typical small brooks and the favorite localities of the fishes. In the area visited by Froehlich the *Apistogramma* species are found almost exclusively in running water, in very many cases even in quite fast running brooks and streamlets. There they live under overhanging grass tufts, in layers of dead leaves or among flooded vegetation. The bottom frequently consists of coarse gravel and stones. The fish then live and spawn in the gaps among the stones. I would refer the reader to the rock or stone aquarium suggested by me earlier. As can be seen, it has been adapted to natural conditions. If such an environment is provided inside a tank measuring 1 m to 1.50 m in length and a strong water current is ensured by means of a powerful rotary pump, the aquarist has made a good job of imitating natural conditions. The water properties will no

Apistogramma macmasteri, female.

↑ *Apistogramma* sp. cf. *macmasteri*.

Apistogramma macmasteri, male. Note the characteristic red marking on the caudal fin. ↓

doubt be of interest, too. Water temperatures recorded by Froehlich within the range of the *Apistogramma* biotope, both in January and in October, lay between 24 and 25, more rarely 26° C. Only *Microgeophagus ramirezi* formed an exception here. The water hardness fluctuated between 0.5 and 2° German hardness, and the pH was around 6.2 to 6.8.

APISTOGRAMMA BORELLII
Borelli's Dwarf Cichlid

Borelli's dwarf cichlid was first imported into Germany in 1938, but only very recently have wild catches appeared on the market again from time to time. The knowledge concerning the native range of

Apistogramma borellii males.

this species had for a long time lain buried and forgotten. J.P. Arnold had stated in 1939/40 that the species was imported by H. Rose (Hamburg) from the central Rio Parana range. Until recently, this quickly forgotten information was replaced with various speculative native ranges. Arnold was right, however. The wild fishes offered nowadays hail from the central Rio Parana range.

As is the case with many other *Apistogramma* species specific characteristics are very difficult to describe, apart from measurements and counts, and the two latter cannot always be relied upon either. The best way to identify adult males of *A. borellii* is by checking the fish against the color photograph. They are round-tailed, small, comparatively stocky fish with an evenly domed frontal profile and magnificently developed dorsal, ventral and anal fins. The dorsal fin in particular—often carried like a sail, high and greatly extended in the caudal direction—is characteristic. It does not have the extended fin tips of the

A pair of spawning *A. borellii.*

"Indian cichlids" such as *A. cacatuoides, A. bitaeniata, A. sweglesi,* etc.

In their reproductive mood the males show a magnificent shiny pale blue in the distal half of the body and the

Above and Below: two aspects of male *Apistogramma borellii.*

Apistogramma sp.

greater part of the dorsal fin. At the same time the head and pectoral region look bright yellow as, in certain places, do the fins. At this stage the cheek stripes characteristic for *Apistogramma* species are absent in the males. The females on the other hand are barely distinguishable from *A. pertensis, A. pleurotaenia,* and other closely related species. Fin development is much weaker than in the males. Adult males can grow to 8 cm in length, but these are rare exceptions. Normally they attain a length of 6 cm. The females, about 4 cm, remain noticeably smaller.

A. borellii is one of the least difficult species of the genus to spawn. These fish can be spawned at a water hardness between 10 and 12 German degrees. But in this case one should not forget to add a disinfectant to the breeding water after spawning for the prevention of egg fungus. More favorable, of course, is a breeding water with a hardness of 5 German degrees maximum.

For egg-laying, the breeding-tank should be provided with an artificial cave, despite occasional reports that the fishes will spawn on a slanted stone if necessary or even on a glass pane of the aquarium. Usually without assistance from the male, the female cleans the chosen spawning place and then attaches the eggs to it by means of the ovipositor. On an average, about 50 to 70 eggs are produced. The male fertilizes the clutch at intervals of several minutes by fanning spermatozoa onto the eggs with powerful fin movements. Often it does not bother to enter the cave before doing this.

Usually the mother cares for eggs and young without help from the male. The latter should, therefore, be netted out if the breeding tank is a smallish one. In bigger

Apistogramma luelingi.

aquaria, however, it can quite frequently be observed that, after initial resistance, the male is tolerated by the female after all. In other words, all transitional stages from "mother family" to "two-parent family" can occur here. Incidentally, the male helping the female to care for the brood need not necessarily be faithful. He may well disappear in order to spawn with another female and then return to his first wife and continue to do his paternal duties.

Whatever happens, the eggs and the fry should be left with the mother or with both parents. There can be no doubt that *Apistogramma borellii* mothers are the best "nurses" in their genus. At 24°C the fry hatch after about 48 hours. Afterwards they are placed in sand holes. Frequently the parent(s) may also move them from one sand hole to another for no apparent reason. Seven to nine days after egg-laying the young are free swimming.

A female *A. caetei* ready to lead her brood of fry. At this stage she displays well-marked spots on the body and facial bars.

Upper photo: *Apistogramma gossei*, male. **Lower photo:** *Apistogramma eunotus?*

APISTOGRAMMA EUNOTUS
High-backed Dwarf Cichlid

This species from the Rio Ucayali and its drainage area in Peru, attains a length of up to 8.5 cm. It is not very demanding with respect to water parameters. These fish are not always easy to identify. The longitudinal stripe on the body ends just before a fleck, which is about twice as high as it is wide, on the base of the tail fin. This black marking, however, either cannot be discerned or can be distinguished only with difficulty, particularly in the male, but at times also in the female. There are related species that exhibit similar markings. The tail is rounded and may have an orange tint in both sexes. In the male, and in the female as well but less strongly marked, typical fleck markings are found in the rear section of the anal fin close to the body. The males, which grow to a larger size, have more of a blue sheen and more strongly developed fins than the females.

The food requirements of *Apistogramma eunotus* are quite modest. Even though the fish readily eat prepared foods, occasional offerings of live food are recommended. This species can also be bred in water of medium hardness as well as in slightly alkaline water. As in the other *Apistogramma* species, here it is also a question of a cave spawner. The number of eggs per clutch varies between about 50 and 100.

Apistogramma eunotus.

APISTOGRAMMA HIPPOLYTAE

Two-spot Dwarf Cichlid

As a typical distinguishing mark this species has a large saddle-like marking in the area of the ninth dorsal fin ray, which extends upward as far as the base of the dorsal fin. This saddle is almost always found in both sexes. In contrast, a thin longitudinal stripe is found only rarely. A spot of about the same size is found on the base of the tail. The tail fin is round in both

Apistogramma hippolytae, male in breeding coloration.

sexes and contains a series of vertical stripes.

The sexes are, for the most part, very similar. Fairly old males occasionally have a slightly docked tail. Additionally, the males grow larger and exhibit a blue sheen, which the female lacks. The male's fins are more elongated, which is particularly conspicuous in the ventral fins.

The fish occur in the central Rio Negro region in Brazil. In the aquarium they are peaceful and are not destructive to plants. They like thickly planted, not-too-light aquaria that are equipped with caves. Neutral and medium-hard water suffices for the keeping in the aquarium. It is possible to spawn a pair in fairly small aquaria. In larger aquaria, you can keep two or three females with each male. The water should exhibit a pH of about 6.5 and a hardness of not more than 12° DH. With up to 200 eggs, the clutches are quite sizeable. After the fry become free-swimming, the fathers often take part in the rearing of the offspring.

APISTOGRAMMA INIRIDAE
Colombian Dwarf Cichlid

These slender, elegant looking fish with the species-typical broad longitudinal body stripe unfortunately do not turn up frequently on the market. Males reach a total length of about 7.5 cm; females remain about three cm smaller. Males possess an extremely high dorsal fin, which is marked with bands, is rounded and is about twice as long as the head. In the males, the other fins are also elongated. When excited, the fish often exhibit a dark-black pattern of spots directly under the longitudinal body stripe.

The homeland of this fish is Colombia, specifically the region of the Rio Inirida. It is found there in decidedly black water.

Apistogramma iniridae.

Apistogramma iniridae is a peaceful species. The fish will thrive only in soft water at pH values of 6 or less. In small aquaria it is possible to keep them in pairs. Males are not very aggressive, even toward one another. Since males are not especially polygamous, one can also keep several pairs in fairly large aquaria. Each pair needs a territory of about 50 by 50 cm. Spawning has been successful so far only in decidedly soft water with pH values of less than 5. Between 80 and 150 eggs are laid. After the brood becomes free-swimming, the male also takes part in the rearing of the fry.

Closely related and easily confused with *Apistogramma iniridae* are the species *Apistogramma pertensis* and *Apistogramma meinkeni.* Both however, can be distinguished from *Apistogramma iniridae* by their longitudinal body stripes, which in *Apistogramma iniridae* extends into the tail fin, and in the two other species end just before the beginning of the base of the tail. Approximately a half or a whole scale later, in the extension of the longitudinal stripe, there follows a tail peduncle spot. In *Apistogramma pertensis* the spot is small and relatively round; in *Apistogramma meinkeni,* on the other hand, it is oval and clearly larger. In height it corresponds roughly to the diameter of the fish's eye. The dorsal fin of *Apistogramma pertensis* resembles that of *Apistogramma iniridae* in shape and size; in *Apistogramma meinkeni* it is lower and does not attain the height of the body.

APISTOGRAMMA CACATUOIDES
Cockatoo Dwarf Cichlid

This attractive species, so easily bred, is familiar to every aquarist.The blunt, short snout lends a somewhat ferocious look to the fish. Characteristic, too, is the doubly pointed tail-fin of the male, the greatly protracted membranes between the anterior spiny rays in the dorsal fin, and the three to four less prominent parallel dark stripes which lie under the main body-band. Often, though by no means always, the upper, and sometimes also the lower, part of the tail fin are adorned with one or more red spots which have a dark margin. The females, too, are readily identified by the parallel bands. The males attain a length of nearly 8 cm. The females grow to only half that size.

The species has been found to occur in two color varieties which perhaps represent geographical races, a less colorful gray form and a strong blue form. The native range of these fishes is the Amazon region.

A large male *A. cacatuoides* displaying to a smaller female.

In a tank of suitable interior, at a water temperature of 24 to 27°C, our *cacatuoides* is a keen breeder. Even in water of medium hardness (15° of German hardness) it is still possible to spawn these fish. They, too, however, are sensitive to water pollution, and that of course includes chemotherapeutic agents. Even anti-fungicides for the protection of the eggs should, initially, be administered in very cautious doses and while keeping the fish under close observation.

The characteristic social structure I described when dealing with *A. trifasciata* is seen in *cacatuoides* as well—one difference being, however, that the territories of the much bigger *A. cacatuoides* females are considerably smaller than those of the female *A. trifasciata.* Sometimes the polygamous male spawns with all the females within just a few hours.

Interesting observations can sometimes be made in the bigger tanks. Occasionally the "harem" includes "females" that never lay eggs. These are young males which have put on the female brood-caring dress and are, therefore, not recognized as rivals by the territorial male. They are quite capable of smuggling cuckoo's

A female *A. cacatuoides* with dark lines below the black lateral band.

eggs into the pasha's nest, i.e. of having a family with one of his females and helping to look after the brood.

This is something we should be aware of when purchasing *Apistogramma* species. "Males in disguise" turn up in other species, too. If we want to buy larger fish and feel sure we can tell the sexes apart we should, wherever possible, add several females to what we definitely know to be a male—not only to meet the social requirements of our fish but

A male *A. cacatuoides* without the obvious spots in the caudal fin.

also to avoid getting only a "male in disguise" instead of the female we wanted.

Easy to spawn though they are these fish leave a lot to be desired where brood-caring is concerned. Often the eggs are devoured for no apparent reason. However, in cases where, after about 48 hours, the larvae have hatched they are looked after very carefully by the parents. They are placed inside depressions in the sand where they continue to be watched over. About 7 to 9 days after spawning the young are able to swim free. At the age of 20 days the shoal of fry breaks up.

APISTOGRAMMA AGASSIZI
Agassiz's Dwarf Cichlid

Earlier mentioning of the "dwarf cichlid" referred to *Apistogramma agassizi.* Later, the application of that name became increasingly more general and fresh attempts were made to find a popular name for this favorite species, for example color-tailed dwarf cichlids or Agassiz's dwarf cichlid. Most commonly, however, the fish is simply referred to as *"agassizi."*

Apistogramma agassizii, variety.

Apistogramma agassizii, variety.

Apistogramma agassizii, female.

Apistogramma agassizii, variety.

Apistogramma agassizii, variety.

Apistogramma agassizii male displaying to a female, trying to attract her into the nest in his territory. Photo by J. Vierke.

An example of a very colorful variety of *A. agassizii.*

The species is native to the Amazon and its southern tributaries in Brazil and Bolivia, but it is also found beyond the water-shed in the upper reaches of the Rio Parana and the Rio Paraguay. There it occurs in slow-running waters—in pools and ponds.

Not surprisingly, considering the huge range of this species, there are a number of subspecies. These differ from one another particularly in the coloration of the males. In some rare individuals, the vertical fins may be bright orange-red in color. Other variations are predominantly yellow on the body while yet others show a shiny blue body-color. It goes without saying that the colors also depend on the light and the emotional state of the animals.

A. agassizi cannot be easily confused with other *Apistogramma* species. Characteristic of the males is the colorful, wedge-shaped extended tail. Often the female shows a more or less pronounced caudal pattern, too. In the female, however, the rounded tail fin is not as protracted as in the male, and the same goes for the anal and, above all, the dorsal fins. With a body length of about 5 cm, the females remain markedly smaller than the males, which attain a maximum length of 8 cm.

For keeping and breeding I would refer the reader to the advice already given when I dealt with other members of the genus. I have to add, however, that this dwarf cichlid is by no means easy to breed. The difficulties are almost as great as with *A. bitaeniata,* even if spawning can succeed at a hardness of 11 German degrees. The eggs are very prone to fungus. *Agassizii* mothers are particularly often seen to fan new water over the brood in a way that is unusual for cichlids. The mother stands just outside the cave entrance, the tail pointing in the direction of the breeding-cave and almost ceaselessly, by means of regular, strong movements of the tail, forces water into the cave.

To make sure it stays put and does not shoot forward the fish must of course counterbalance these tail movements with powerful strokes of the pectoral fins. Eleven days after spawning the fry are free swimming. During the first few days the youngsters are led back to the cave each night by their mother. If the procession is too slow for her she takes the stragglers into her mouth and carries them to the cave where their siblings have already made themselves at home.

Regrettably, the sexes are often unevenly represented among the fry. One either raises a predominantly male generation or an almost exclusively female one. The reasons for this are not clear. One dwarf cichlid breeder recently told me that, if artificially raised, almost all tank-bred fishes develop into females, but if the fry were left with the mother nearly all of them grew into males. Other breeders regard the pH at the time of spawning to be the decisive factor. A low pH is said to result in a surplus of males while a rising pH is held responsible for an excess of females.

APISTOGRAMMA RAMIREZI Also known as *MICROGEOPHAGUS RAMIREZI* Butterfly Cichlid

Apistogramma ramirezi is a small gem—a dwarf cichlid with extremely beautiful colors and very lively yet always peaceful behavior. Anyone who has watched this colorful

A typical aquarium-bred specimen of *Microgeophagus ramirezi* photographed by Dr. Herbert R. Axelrod many years ago.

little fish perform its fluttering courtship dance will confirm the aptness of its popular name.

In some respects *Apistogramma ramirezi* does not fit its genus. There is hardly any sexual dimorphism in this species and the animals are "open breeders" and form "two-parent families." For these reasons many authorities place this fish in a genus of its own and refer to it as *Papiliochromis* or *Microgeophagus ramirezi.* Whether this action is justified is debatable. However, there is no need for the aquarist to adapt to the new name, as generic names are not obligatory to the same extent as are specific names. I shall here continue to use the old classification.

This particularly colorful species with its high build bears black cheek stripes, extended black spines in the dorsal fin, a black spot in the body center, and frequently an extra black spot, or several extra ones, on the body or in the dorsal fin. Presumably the distribution and intensity of the markings depend not only on the emotional state of the fish but also their geographical origin. The basic color can vary a great deal. In the light the scales on the body are a

For spawning the butterfly cichlid it is not necessary to create cave-like arrangement of rocks needed by other *Apistogramma* species. A piece of flat stone or slate will be sufficient.

shiny pale blue. Lately commercial breeders have been producing xanthochromic forms, the golden rams.

It is not always possible to distinguish the males by the more strongly developed spines in the dorsal fin. A reliable sexual characteristic is the visibly red to reddish-violet abdominal region of the females. This makes the gravid females look much more magnificent than their males.

The fish come from northern and northwestern South America—unless, of course they hail from German or Southeast Asian hatcheries. Localities where these fish are found are known from Venezuela and Colombia.

While temperatures within the lower ranges suffice for

Left photo: A male butterfly cichlid waits as the female adds a few more eggs to the nest. **Photo above:** The male (fish on the right) fertilizes the egg.

keeping the species, the water temperature in the breeding-tank should be raised to about 30°C. The water should be soft and slightly acid. Aeration should be provided. Partial water changes at frequent intervals are necessary, too, as the fish are sensitive to nitrite. The species is a typical "open breeder" using rounded stones or, sometimes, depressions in the gravel or stone as its spawning substrate. Male and female usually occupy a common territory and raise the brood together. If one female is placed in a larger tank with several males a different social structure results. Then the males form territories and the female flirts her way through all the males in succession. Whichever love she happens to be with she helps to defend the territory against the neighbors (which are now, of course, regarded with much stronger suspicion). After spawning she frequently changes sides, so that a week later another happy father is presented with a stone full of eggs.

If we intend to breed the fish, they must be given plenty of live food. White and black mosquito larvae as well as *Cyclops,* seem to be particularly effective when it comes to getting the female into the spawning-mood. As one would expect of "open breeders," the number of eggs produced is much higher than it is in cave breeders—150 to 400 eggs being the norm where *A. ramirezi* is concerned. At 30°C the fry hatch after 48 hours. At 25°C hatching is postponed for another day. Now the brood is carried into one of the pits that have been dug in advance for the purpose. The yolk sac of the fry takes an additional 6

This photo illustrates the changeable character of the mid-body spot in the butterfly cichlid during the breeding phases.

days to be resorbed, after which period the young are free swimming and follow the parents. From then on they are easy to raise. They do need plenty of food, however, (with nauplii of *Artemia* being the easiest food to provide) and the water in their tank has to be partially changed at frequent intervals—else growth will be stunted. When the aquarist buys butterfly cichlids he must make sure he is not given stunted fishes already exhibiting their full adult coloration—such fish will show little further growth. A good specimen can be expected to attain a total length of 5 to 6 cm.

The fish in these photos, originally called *Crenicara altispinosa*, is now considered allied with the butterfly dwarf cichlid. A mature male shown below.

APISTOGRAMMA NIJSSENI
Panda Dwarf Cichlid

Although panda have no pronounced fin formations, they can be distinguished immediately from all other *Apistogramma* species that are kept by aquarists at the present time through the typical coloration of both sexes.

The gleaming blue males grow a good two cm longer than the females, which reach a length of only about four cm. Males possess elongated ventral fins and have a semicircular yellow to bright red border stripe on the tail fin. The red coloration of this edging apparently can be positively influenced by the

Apistogramma nijsseni, a much smaller female and a more colorful male.

During sexual display a male *A. nijsseni* can develop a well marked pattern; a female develops the reverse of that pattern.

Apistogramma nijsseni, an adult male bred in the aquarium.

careful addition of iron (values of more than 0.05 ppm Fe may be harmful!). By way of suggestion, this ring can also be seen in the female.

Females are provided with conspicuously large black spots on a yellow ground color in the gill cover area, at the start of the dorsal fin, approximately in the middle of the body, and on the base of the tail. The ventral fins of the female are also a deep-black color.

This fish comes from the region of the lower Rio Ucayali in Peru at Jenaro Herera. In it the aquarist has a peaceful *Apistogramma* species that does not destroy plants and which can also be kept in medium-hard, neutral, or even slightly alkaline water. It should be kept in dimly lit aquaria that are well planted and provided with caves.

Close-up of a female panda dwarf cichlid at the height of the guarding coloration.

Since the males are more or less monogamous, spawning in pairs should be possible. Nevertheless, frequently breeding seems to present problems.

The males take part in leading and caring for the free-swimming brood, and at those times exhibit a coloration that strongly resembles that of the fry-leading female. Their blue sheen becomes pale and makes way for a yellow tone. On the body the black spotted panda coloration of the female forms. The spot in the middle of the body, however, remains conspicuously pale, as it does in the fry-leading mother. Unfortunately, the sex ratio of the offspring is often extremely unbalanced.

Etroplus maculatus is a well-known aquarium fish.

ETROPLUS MACULATUS
Orange Chromide

Besides a few *Tilapia* species, the only additional cichlid genus that occurs in Asia is the genus *Etroplus.* To this genus belong the three species: *Etroplus canarensis, Etroplus suratensis,* and *Etroplus maculatus* from southern India and Sri Lanka. The glaring-yellow-colored *Etroplus canarensis* is still completely unknown to aquarists. *Etroplus suratensis* reaches a total length of about 40 cm, and thus grows far beyond the domain of the dwarf cichlids. On the other hand, one can designate *Etroplus maculatus* as a dwarf cichlid without reservation. The fish remain small, do not destroy plants, and are absolutely well-mannered toward other species.

Etroplus maculatus, is a peaceful and undemanding fish, which is also suitable for the community tank. In addition to the wild form, a gold variety, which lacks the black markings of the wild form, has also been on the market for a number of years. However, the wild form also exhibits a beautiful yellow coloration, from time to time.

Etroplus suratensis, a species of large dimension that is not as popular as the orange chromide in the hobby.

The sexes can be distinguished only with great difficulty, at least in dealers' tanks. Females of the same age are, however, smaller and more compressed than the males. During the spawning period, the males are rich canary yellow.

Breeding these Sri Lankans is so interesting that no dwarf-cichlid fancier should pass up the opportunity to do so. When the fish are in breeding condition, the male and female choose a territory, which they defend together against other fishes. Now they begin to dig depressions in the sand, usually at the base of fairly large stones. Soon after, the fish inspect various possible spawning sites such as diagonally standing stone slabs, large flowerpots, aquarium panes. They show a preference for all stable, vertical surfaces. Again and again they suck on the stone that happens to interest them at the moment; it often appears as if they want to bite off a piece of rock!

Soon the fish glide with their vent over the spawning substrate simultaneously, or one after the other. In the meantime one can see a genital papilla in both sexes. Finally, the female sticks the first eggs on the substrate. The male subsequently glides with propeller-like pectoral fin movements over the eggs and fertilizes the clutch. These events are repeated again and again, until the clutch, consisting of 100 or more eggs, is complete. The eggs are camouflaged and hang on tiny stalks.

After spawning, one mate constantly stays over the clutch and moves a stream of fresh water over the eggs by fanning the pectoral fins. Meanwhile, the other mate inspects the territory or occupies itself with the digging of new holes.

The young *Etroplus* hatch after about three days. The parents immediately gather the fry at the base of the spawning

A yellow or xanthic form of *E. maculatus*.

A breeding pair of orange chromides will not hesitate to spawn on a vertical surface.

stone in a hollow, or they carry them to one of the newly dug depressions. Here the brood lies as a swarming, brown-gray mass, always carefully guarded by the parents. From time to time the youngsters are transferred to another hollow. This is sensible, for in this way it is ensured that the little ones always lie on clean substrate.

The young *Etroplus* do not become free-swimming until about the sixth day after hatching. In a dense school, they stay by one of their parents, while the other parent keeps away possible predators. If some of the fry happen to stray too far away from the leading adult, this parent jerks several times with the now deep-black ventral fins, and the band of fry immediately returns!

From the start, the young *Etroplus* eat the smallest infusoria and freshly hatched brine shrimp. This food, however, is not sufficient for optimal rearing. A close look will show that the youngsters now and then literally rub against their parents. This behavior causes the adults' dermal glands to produce more slime than normal. This slime is nutritious and serves as an important source of food for the fry.

As in orange chromides, a male and a female *E. suratensis* are indistinguishable externally during the non-breeding stage.

In order to prove this, fry-leading *Etroplus* parents were removed from the tank, rolled briefly in coal dust, and returned to the fry. Soon the fry became accustomed to the now black parents. They again began to rub against them and to eat from them. In subsequent stomach examinations of the fry, the presence of coal could be established. This was the proof! Without this slime food, the fry suffer and many die prematurely.

The genus *APISTOGRAMMOIDES*

The only known representative of this genus is the *A. pucallpaensis* from Peru. Unlike members of the genus *Apistogramma,* whose development would appear more advanced, it does not have three or four spines in the anal fin but eight. This characteristic is easily seen in the living fish, too. Equally typical are three prominent dark spots, one above the other, which cover the whole height of the tail-fin base and are surrounded by a narrow halo of shining gold. The dorsal fin is low and, like the rest of the fins, comparatively short. The caudal fin is rounded. The species is only rarely imported. It requires soft water. Keeping and breeding are the same as for the genus *Apistogramma.*

This page: Shown are individuals of *Apistogrammoides pucallpaensis* showing some variations in body pattern. However, note the typical tail spots present in both fishes.

The genus *TAENIACARA*

From the central Amazon hails the species *Taeniacara candidi,* so far the only known representative of its genus. The fish are extremely slender and elongate. Similar to Agassiz's dwarf cichlid, the caudal fin is wedge-shaped and protracted. The greatly extended ventral fins of the males have orange-colored tips. Their dorsal fin, on the other hand, remains low. These dwarf cichlids attain a maximum size of about five centimeters. They are only seldom imported and breeding is not easy.

Taenicara candidi. The extremely slender body is very evident in this fish.

The genus *CRENICARA*

These attractive little fishes are native to the central Amazon range. While it is possible to keep them in water of medium hardness, they prefer water with a hardness of less than 10 German degrees. Anyone wanting to try and breed them should use softer and slightly acid water, if possible.

Of the three species so far described *C. punctulata,* with a total length of a good 10 cm, is

Crenicara maculata, female.

the largest. Like *Crenicara maculata,* which grows to no more than 8 cm in length, this fish only seldom appears on the market. "*C. praetoriusi*" would appear to be a synonym of *C. maculata.* The third

Crenicara punctulata.

species, *C. filamentosa*, on the other hand, is more readily available, being accidentally caught together with the cardinal tetra, *Cheirodon axelrodi.* Adult males in their full colors are among the most beautiful dwarf cichlids. All species show characteristic markings formed by two rows of quadratic dark spots arranged like a chessboard. This explains their popular name, checkerboard cichlids.

Crenicara filamentosa, female. Note the cherry red ventral fins and checkerboard-like markings.

CRENICARA FILAMENTOSA
Fork-Tailed Checkerboard Cichlid.

Males of this species of slender little fish with conspicuously blunt snouts are characterized above all by the greatly protracted bilobate tail-fin and their shiny red, blue, and black fins. With a maximum length of 6 cm, the females remain considerably smaller than the males which grow to a good 9 cm. Generally speaking they show a black and white chessboard-type pattern on the body. Their fins are perfectly translucent. In the gravid female, however, the ventral fins turn reddish and in brood-caring females

C. filamentosa, male. Note the forked tail that is developed in the males only.

they become bright cherry-red.

If the aquarist is already able to differentiate the sexes—forked tail or reddish ventral fins—when he makes his purchase, he is advised to acquire several females to one male if possible. As regards its social structure, this species is similar to members of the genus *Apistogramma.* In a species tank the fish tend to be rather shy, and again it is an advantage to associate them with characins or live-bearers. Checkerboard cichlids love their tanks to be densely planted and their favorite water temperature lies around 25°C. Breeding caves can be dispensed with; they are rarely made use of by these fishes. Usually the female sticks the eggs on a leaf which is standing in the horizontal position or a flat stone.

One clutch comprises about 60 to 120 eggs. Afterwards the mother, her head threateningly lowered, chases away every tank-dweller who comes near the brood, including the father. Unfortunately the spawn is not seldom devoured, and if soft water is not available to use, the eggs become fungused. The eggs hatch after about 60 hours. They are not deposited in sand pits but left on the topside of the leaf. A further 100 hours elapse before the young swim free. The mother continues to look after the shoal of fry for a few more weeks.

The genus *NANNACARA*

In their behavior the few species that make up this genus are very similar to *Apistogramma* species. In these South Americans, too, we come across the territorial set-up I have already described, i.e. a "super-territory" for the male and inside this several sub-territories for the females. We should take this into consideration when we stock the tank—several males to one tank often means an unhappy

A golden dwarf acara male waiting for his turn to fertilize the eggs being laid by the female.

Nannacara anomala, a male in breeding coloration.

ending. Certainly only one of the males manages to establish territory, as a result of which the others are more or less permanently on the run. These experiences were made even in tanks with a length of two meters and a capacity of 320 liters!

Nannacara anomala
Golden Dwarf Acara

The casual observer seeing these fishes in the dealer's tank will have little or no idea just how beautiful and interesting these dwarf cichlids are. They make no special demands upon the composition of the water. In water of medium hardness, which should be slightly acid if possible, they are content, too. Whenever possible, we should give them live food, although dried food will be accepted if necessary.

To describe the coloration of the males is virtually impossible as there is sheer endless variation in markings and colors, depending as they do on mood and, presumably, on the origin of the fish as well. The edges of the scales usually have a metallic green or bluish sheen while in the center the scales show a small triangular spot.

The colorless females remain much smaller than the males, the latter reaching a length of about 9 centimeters. In the females, too, the markings can change with surpisingsuddenness.

Nannacara anomala originates from Guyana. The species is so easy to breed that it can be recommended to the beginner. Individual pairs can be placed in well-planted 50- to 70-liter tanks with soft or medium-hard water of about 25°C. Very soon the male will start his tireless courtship of the female. There is no actual pair-formation, however, until the female is ready to spawn—when it loses its dark markings, takes on a dirty-brown coloration and follows the courting male. At this stage the pair begin to clean flat stones or caves in various parts of the tank with quick movements of the snout. Not much later cleaning is confined to a single spot—the chosen spawning place. Eventually the female lays her eggs, attaching them with the ovipositor. During the short intervals between egg-laying the male glides over the eggs and fertilizes them. Spawning is completed within about half an hour, by which time the 50 to 300 eggs have been neatly arranged in a circular clutch.

As soon as the last egg has been laid and fertilized the parnership comes to an end. Now we have to net the male out of the breeding tank, or else it would be chased without mercy or, as not infrequently happens, even killed by the smaller female, former husband or no.

While still engaged in spawning the female puts on its brood-caring dress—2 to 3 rows of coarse, light colored spots on a dark base. After about 48 hours the fry hatch with their yolk sacs, are sucked out of the egg membrane by the mother, and carried into a pit.

After another 5 days the little ones are free swimming and form a dense shoal which is now led by the mother. Experiments carried out by

Kuenzer with inanimate objects proved that the following-the-mother behavior of the fry is triggered off by the black-and-white coloration of the female and her jerky way of swimming.

However, to generalize on the family life of *N. anomala* from these particular experiences of breeding behavior would be a mistake. In large tanks the male takes on the outer territory and so guards his brood indirectly at first. Often, when the young have been swimming free for a few days, the father also looks after a portion of the fry directly. His body then shows black and white markings similar to those displayed by the female—obviously, if the fry are to follow him! Sometimes the two adults actually fight over the care of the free-swimming brood. In such cases, as distinct from the earlier fights immediately after spawning, it is invariably the father who emerges victorious.

Two *Nannacara anomala* males in a tug-o-war, testing each other's strength. This can occur at any phase of the breeding cycle.

Nannacara aureocephalus, a female guarding her fry resting on the vertical surface of a rock.

NANNACARA AUREOCEPHALUS
Gold-headed Nannacara

This species, which was described in 1983, is quite similar to *Nannacara anomala,* but grows larger. For spawning, soft water and a pH value of between 5 and 6 is needed. The females of the two closely related species are practically impossible to differentiate, but the males clearly are differently colored. The surest character is the color mode of the body scales in fully colored specimens. In *Nannacara aureocephalus* the scales have dark edges and have their glossy zone on the inside. In *Nannacara anomala* the arrangement is exactly reversed.

The genus *AEQUIDENS*

One member of the group of *Aequidens* species, a truly charming little representative, is considered to belong with the dwarf cichlids, *Aequidens curviceps,* the flag cichlid. Although sometimes reaching a good 10 centimeters in length in their natural environment, they only seldom exceed 7 cm in total length when kept in the aquarium.

The fish form a typical two-parent family and consequently show (almost) no sexual dimorphism. This means we have a problem when it comes to finding a pair of breeders. Among adult fishes the males can often be identified by their more

A male *Aequidens curviceps* waiting eagerly for his turn to fertilize the eggs his partner started to lay.

strongly developed dorsal and anal fins. But the females, which usually remain a little shorter, often have pointed fins as well.

Although originating from the Amazon region, the flag cichlid can without reservation be kept and bred in water of medium hardness, too. The water should, however, be partially replaced at regular intervals, as in old water the fish are highly susceptible to disease.

These cichlids are relatively easy to propagate even in fairly small aquaria. As with all cichlids, a good, varied diet of live food helps to get the fish into a "reproductive mood." The water temperature can be anything between 20 and 30°C, although the most favorable temperatures appear to lie within the median range, i.e. between 23 and 27°C. Even in the breeding season the flag cichlids virtually never dig up any plants. This makes them safe inhabitants of attractively planted community tanks—all the more so as they are very peaceful among themselves and towards other species. In fact, the flag cichlid is probably the most peaceful

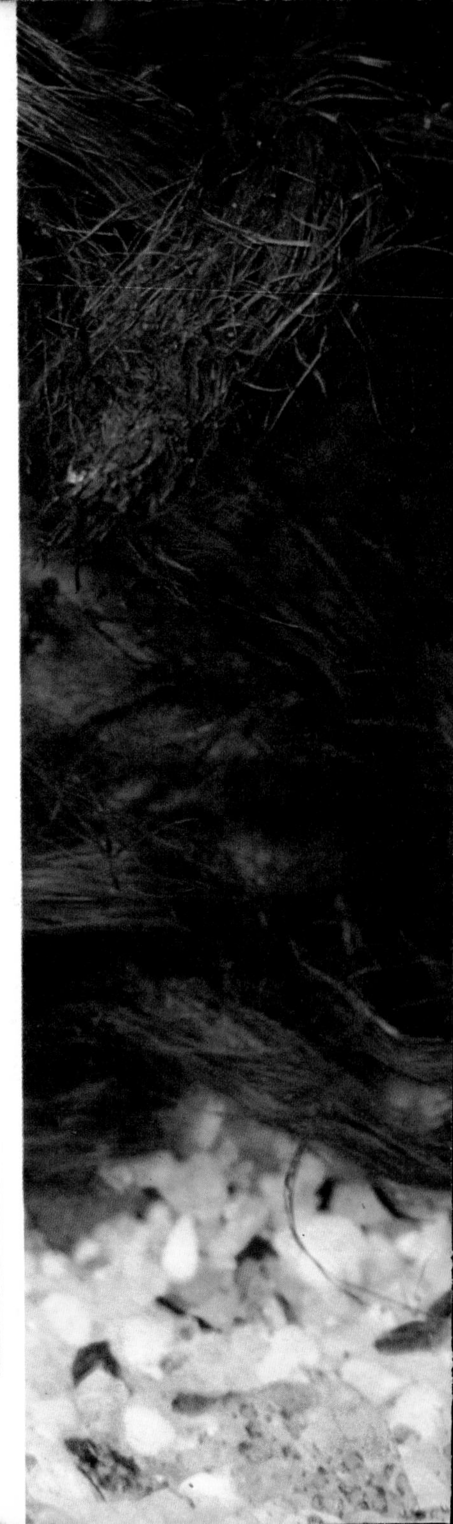

This flag cichlid is guarding the fry that were transferred from the spawning site to a pit dug on the substrate.

The markings of male and female flag cichlids are almost alike. However, during the breeding period the sexes are identifiable.

cichlid we know. The usual spawning substrate selected by these fish consists of a flat stone or, sometimes, the root of a tree. A clutch generally comprises up to 300 eggs, although particularly large females can occasionally produce three times that quantity.

When the fry have hatched they are carried into pits and guarded until they are ready to swim free. Immediately after they have begun to swim free we give them their first food consisting of newly-hatched nauplii of *Artemia.* The aquarist should not lose heart if the parents eat the first brood or two. Not much later they invariably prove to be really excellent at the task of broodcaring. It would be a pity to raise this species artificially, not only because we would cheat ourselves out of that touching *curviceps* family scene, but also because we would be producing behaviorally degenerate, egg-eating strains.

AEQUIDENS DORSIGER
Red-breasted Dwarf Cichlid

In addition to *Aequidens curviceps,* one also frequently finds its close relative, the even more splendidly colored *Aequidens dorsiger,* in aquaria. It attains a maximum total length of 8 cm and prefers soft or medium-hard water of about 24° C. Both species resemble each other in body form and behavior. *Aequidens dorsiger* has a dark longitudinal stripe which extends from the rear edge of the eye to the middle of the body and which usually ends there in a fleck. In addition, a dark fleck, usually surrounded by a light region, is often present in the dorsal fin. In dealers' tanks the fish are not very attractive, but at breeding time one must—particularly because of their exquisite red-colored throat and breast area—number them among the most beautiful dwarf cichlids! There are, however, also races whose red tints are more inclined to blackish.

An adult male *Aequidens dorsiger* develops the characteristic red breast. He has already cleaned the substrate, a flat stone below, where the female will lay eggs.

Now to the requirements for keeping this species of dwarf cichlid, which is found in the marshy regions of the Rio Paraguay. It is a decidedly peaceful dwarf cichlid, which makes no particular demands with respect to water or food. The water hardness can be up to 20° DH, but softer water is better. For a pair, aquaria that are 60 cm on a side, or larger, are sufficient. They are excellent for the well-planted community tank. A sand substrate, several smooth rocks, dense planting on the edges, and dug out areas under roots should not be lacking.

For spawning, one should keep the fish in pairs and feed them abundant amounts of live food. Females lay up to 500 eggs (normally about 200), preferably on the top side of smooth pebbles, but occasionally on vertical substrates or on the sand. The eggs are transparent amber. The eggs hatch after about 48 hours. They are then placed by the parents in small hollows in the substrate for the next five days where they are guarded.

As soon as a batch of eggs are laid, this male red-breasted dwarf cichlid waiting on the side will certainly move in and release sperm on the eggs.

After they become free-swimming they should be fed freshly hatched brine shrimp. The youngsters grow rapidlly with regular partial water changes, but are, however, still quite sensitive to changes in water chemistry in the first 4 weeks.

The genus *PELVICACHROMIS*

Occurring above all in tropical western and central Africa are a few cichlids of smallish to medium size which were once considered to belong to the genus *Pelmatochromis.* Some of these may be described as true dwarf cichlids. Repeated revision of this group of fishes resulted in a multitude of new names for science and, for us, the necessity to re-learn them.

Most small cichlids from the one-time genus *Pelmatochromis* were, in 1971, transferred to the new genus *Pelvicachromis.* The most important of these are: *P. humilis,* Sierra Leone to southeastern Guinea, up to 12.5 cm; *P. pulcher* (previously *kribensis, aureocephalus, camerunensis*), southern Nigeria, up to 10 cm (although in nature very occasionally exceeding that length); *P. roloffi,* Sierra Leone, Guinea, Liberia, 9 cm maximum; *P. taeniatus* (in the past often called *klugei* or *kribensis*), southwestern Cameroon to Nigeria, about 8 cm; *P. subocellatus,* Congo to Gabon, about 10 cm.

The best-known representative of the newly established genus *Chromidotilapia* is *C. guentheri,* a quiet, peaceable mouthbrooder which, however, attains a size of up to 18 cm. The other members of this genus, too, become so large that they can no longer be regarded as dwarf cichlids. One exception that may be mentioned here is *C. schoutedeni*, a fish from the Congo region. So far, however, hardly anything is known about this species.

Thysia ansorgei is another fish that used to be classified as a member of the genus *Pelmatochromis.* It was also referred to as *P. arnoldi* or *P. annectens.* The species is peaceful, but not particularly attractive-looking. Because of that, and not least of all the fact that it grows to a total length of 13 cm which makes it too big to be considered a true dwarf cichlid in any case, this brief mention of the fish shall suffice.

One dwarf cichlid that must not be forgotten here is the "African butterfly cichlid," whose name *Pelmatochromis thomasi* now is changed to *Anomalochromis thomasi.*

PELVICACHROMIS PULCHER

Kribensis

This species, in Germany also known by the popular names "King Cichlid" and "magnificent purple cichlid," fully deserves its Latin name "beautiful." In the distal part of the dorsal fin and the upper half of the caudal fin the fish often, though not always, have black spots with a yellow margin—so-called "eye-spots." Depending on their origin, the fish vary in appearance. Perhaps some forms still need to be taken out of the species *P. pulcher.*

A male kribensis.

A female kribensis with the characteristic deep purple belly.

The "kribensis" hail from the lower range of the Niger and Kribi Rivers. Often they even penetrate into the brackish zones. This means the water hardness is only of minor importance. Within their area of distribution they show a preference for the ranges of more powerful rivers. There, however, they favor the river edges with their dense growth of submerged plants where the current is greatly diminished.

Above: *Pelvicachromis pulcher,* male. **Below:** Developing eggs are guarded carefully by the female kribensis. During this period she is very territorial and aggressive.

Here they turn up the bottom layer and catch small prey. The fish build their own breeding caves—again inside the underwater jungle—by removing sand from underneath a stone or a branch. They then attach their eggs to the ceiling of the cave.

In its natural environment *P. pulcher* occupies a breeding territory of about ¼ m^2. By cichlid standards this is comparatively small. Nevertheless it should make one a little thoughtful when looking at such limited space as we are able to offer in our aquaria.

In smallish tanks it is advisable to keep them in pairs. Gravid females are readily identified by the prominently red, swollen belly and the more or less rounded dorsal and anal fins. The tank should be well planted and divided, and contain some caves. Then the digging will, in the great majority of cases, be confined to the area of the breeding cave.

Once the 200 to 300 eggs have been laid it is in the father's own best interest to keep away from the breeding cave, else the mother turns vicious. In tanks which are too small we sometimes have to save him from the aggression of his spouse. In a tank of sufficient size, however, he takes it upon himself to defend the territorial boundaries. When the young are free swimming, after 8 to 9 days, the father usually takes a completely equal part in the care and leading of his offspring. As soon as the fry

are free swimming we have to offer them newly-hatched nauplii of *Artemia.* With a good varied diet the new generation may already be sexually mature after 7 to 8 months. Often they are kept at too high a temperature—optimal temperatures lie aroud 24°C.

Facing page: This female kribensis is laying eggs on the ceiling of the spawning cave.

Pelvicachromis taeniatus, male. Note the wire-net-like pattern on the body scales and the distinctive caudal spot.

PELVICACHROMIS TAENIATUS
Striped African Dwarf Cichlid

The popular name "striped cichlid" is somewhat misleading. This cichlid, too, is represented by a number of geographical races. The best-known of them is the form originating from southwestern Cameroon, from the catchment area of the river Lobe. The males of this subspecies do not have spots in the caudal fin. Another race comes from the river Kienke in central western Cameroon. In the upper part of its caudal fin, where the edge is bright-red, it possesses prominent black eye-spots with a golden margin.

Similar spots in the caudal

fin, and in addition a pattern of longitudinal striations in its lower portions, are also shown by the variation from western Cameroon that used to be known as *P. klugei.* In the males the body scales are particularly dark along the edges so that the body looks as though it were covered with wire netting. The females can be identified by their greenish-blue to violet-blue bellies.

Reproduction is more or less the same as for *P. pulcher.* One clutch consists of 40 to 150 eggs. The fishes are keen spawners. Raising this species, unfortunately, is often beset with problems and not as easy as with *P. pulcher.* The fry are very sensitive to infusioria. But the experiences that have occurred with this fish vary, apparently depending entirely on where the fishes come from. Some hail from brackish river estuaries where they occur in association with *P. pulcher.* Others originate from small jungle brooks.

Facing page: Two views of a pair of striped African dwarf cichlids spawning on the wall of a coconut shell. Note the bulging breeding tube of the female fish in upper photo.

PELVICACHROMIS SUBOCELLATUS
Eye-Spot Cichlid

P. subocellatus is native to equatorial Africa from Gabon to the lower Congo. It is more compact in build than the species described earlier and lives in estuaries where it penetrates into the brackish zone. It is also found in the lagoons, not far from the coast, which sometimes carry fresh and sometimes brackish water. Unfortunately this very attractive, easily cared for cichlid is only offered fairly sporadically. Particularly beautiful are the courtship colors of the females, the anterior and posterior third of the body being almost black at this time while the middle third forms a light-colored "saddle". The upper zone of the saddle and the part of the dorsal fin it is in contact with shine silvery-white. The lower half and the ventral fins are bright red. Apart from this the coloration of these fishes is so variable that a precise description would be pointless. The only other characteristic I would like to draw attention to is the eye-spot present in the female—it adorns the posterior portion of the dorsal fin. It is this the Latin

A male *Pelvicachromis roloffi* guarding his "house," one-half of a coconut shell. Note the absence of the purple color present in the females.

One of the many color forms of aquarium-bred *Pelvicachromis subocellatus.* However, the red belly that appears in many species of *Pelvicachromis* persists.

specific name is referring to.

Keeping and breeding are the same as for *P. pulcher.* If wild individuals are difficult to acclimatize and show no interest in food, an addition of sodium chloride at the rate of up to 7 g of salt per liter is recommended.

From southern Nigeria come variations that are very similar to *P. subocellatus.* Whether these fish are special races of this species or variants of another species not yet described still remains to be established.

PELVICACHROMIS ROLOFFI
Golden Cichlid

In this species, which occurs in Sierra Leone and the bordering areas of Liberia and

Guinea, it is again the female which is the more colorful of the two sexes. It is golden-yellow in color and has a deep-violet abdominal region. In the tail-fin and at the base of the dorsal fin one often finds one or more eye-spots.

This uncommon species is much more difficult to keep and breed than the *Pelvicachromis* species discussed above. It can, therefore, only be recommended to experienced aquarists.

ANOMALOCHROMIS THOMASI

African Butterfly Cichlid

Just as the butterfly cichlid, *Apistogrammi ramirezi,* stands out from the circle of *Apistogramma* species by its appearance and behavior, so *Anomalochromis thomasi* from Sierra Leone, southeastern Guinea and western Liberia differs from the other members of its former genus. To this is added the fact that in shape, movement, and broodcaring behavior it bears such a strong resemblance to its South American counterpart that its popular name "African Butterfly Cichlid" is very apt indeed. This resemblance

A golden cichlid female in spawning colors.

does not mean, however, that the two fishes are related in a systematic sense.

The first spines in the dorsal fin of *M. ramirezi* are jet black in color and frequently extended. This characteristic is absent in *A. thomasi.* Furthermore, the African does not have the cheek stripe pointing towards the back as in *Apistogramma* species. In *A. thomasi* the cheek stripe points towards the front.

Sexing is difficult until the fishes form pairs. With a length of about 7 cm the female remains smaller than its partner which can attain a maximum size of 10 cm. When the fishes pair off the females already have markedly rounded bellies. Soon the fish

Anomalochromis thomasi. The sexes are very similar, but the females have rounder bellies when gravid.

start to clean stones and roots in their territory. The female does this with particular zealousness. Finally only the chosen spawning stone is still being cleaned, and then the eggs are attached to the substrate and fertilized. About half an hour later the clutch is complete, consisting of 300 to 400 eggs. Now the parents take turns looking after the spawn.

A flat substrate is utilized by this spawning pair of the African butterfly cichlid.

After about 48 hours the embryos start to move inside the egg membranes. The parents help the young to hatch and carry them to the pits already dug for this purpose the previous day. After a period of about four to five days the fry have resorbed their yolk sacs, rise above the pit, form a shoal, and allow the parents to lead them.

The keeping and breeding of this appealing and peaceful species present no problems. The fish can, therefore, be recommended to inexperienced aquarists and are suitable for the community tank. As a guide for the water temperature I would suggest 25°C. The other water values are unimportant.

The genus *NANOCHROMIS*

In this genus are united a number of very slender little fish which are closely related to the *Pelmatochromis/ Pelvicachromis* group. At the time of writing only two species, *N. nudiceps,* the Congo dwarf cichlid, and *N. dimidatus,* the dimidiatus, can be found in our tanks. Of the remaining 12 species it is very, very rare to find them in pet shops.

The Congo dwarf cichlid and the dimidiatus should be kept in soft water if possible, although a hardness of up to 20 German degress will be tolerated. The most favorable temperatures lie around 25°C.

The propogation of these fishes is not easy. Sometimes it succeeds in small tanks. What matters most of all is that the aquarium is well stocked with plants and contains a cave. Problems start when the pair are put into the breeding tank. All too often given the chance they will fight each other to the death. If necessary, the fish should be separated by a glass pane (placed inside the tank) for a few days. Thereafter it will usually be safe to let them into each other's company again.

The two species spawn in caves which they prefer to excavate themselves. Their 60 to 80 (in exceptional cases up to 250, it has been said) yellowish to reddish eggs are suspended from the ceiling of the cave by short threads. In small tanks the two partners generally start to fight when spawning is over. Then the male must be netted out at once, else it will get killed. In an aquarium of adequate size the father helps to raise the brood once they are free swimming.

The fry hatch after three to four days and start to swim free after another three days. Whatever happens, it is vital to leave the female in the tank with them. Attempts to incubate the eggs artificially have so far almost always failed.

The Congo is also the home of a few other cichlids which, while not considered to belong to the genus *Nanochromis,* are very interesting too and equally adapted to the very strong water current. *Steatocranus casuarius,* with the imposing hump on its head, *Lamprologus congoensis, Leptotilapia tinanti,* with its gigantic shovel-mouth, and the dark-brown *Teleogramma brichardi,* are among the best-known of them. All of them are very slender bottom-dwellers which lay their eggs in caves and rock crevices.

NANOCHROMIS NUDICEPS
Congo Dwarf Cichlid

Fish of this species are caught in the vicinity of Kinshasa. The males reach a total length of about 8 cm; the females do not exceed 6 cm in length. Their long, slim bodies indicate that they are adapted to fast-flowing waters. The pale-gray or whitish-yellow fish only reveal the shiny bluish-

A photograph of a pair of Congo cichlids taken when early commercial importation of this species was taking place.

green or violet scales on the body in the appropriate illumination. Gravid females have a misshapen-looking swollen belly while the males with their sunken abdomen look positively starved. In the sexually mature female the genital papilla also protrudes during the breeding season. Often the male's genital papilla is visibly protruding as well.

NANOCHROMIS DIMIDIATUS
The Dimidiatus

The dimidiatus lives in the Ubangi, a tributary of the Congo. In appearance and behavior it bears a greater resemblance to some *Pelvicachromis* species (e.g. *taeniatus*) than to its blue relative. The fish attain a length of about 7 cm, although the males may exceed this by an extra cm.

Depending on their emotional state the basic color of these fish varies from gray to orange. The colors in the abdominal and pectoral region are particularly intense. In the male, during courtship, these

N. dimidiatus parent raises the fry for a certain period of time. With growth the yolk sacs of these fry will gradually become smaller and disappear.

areas can be bright red. In courting females the belly is of a conspicuous lilac-red color. Another beautiful adornment of the females is their dorsal fin with its silvery-green and whitish edge. In contrast to *N. nudiceps* the genital papilla does not appear until a few hours before spawning.

Facing page: (Upper photo) According to the author this male fish is probably an undescribed species closely related to *N. dimidiatus.* (Lower photo) *N. nudiceps* tending the eggs.

An identified specimen of *Nanochromis.*

A pair of *Nanochromis dimidiatus,* the female on the right.

A pair of *Nanochromis* species. They will, like other species of the genus, probably spawn inside the coconut shell, too.

The genus *PSEUDOCRENILABRUS*

This genus embraces some very undemanding and interesting little mouthbrooders from Africa. They make no special demands either on food or on water quality. They are comparatively peaceful fish although they may do some occasional digging. During the breeding season they prepare pits in the sand for spawning. Most of the time their burrowing activities remain within tolerable limits, however, even in the community tank.

Originally the species of this genus, sometimes also referred to as *Hemihaplochromis,* belonged to the large genus *Haplochromis.* Today the fishes described as *Pseudocrenilabrus* have been separated from those bearing characteristic round, yellow or red, spots (egg mimicry) on the anal fins. These *Haplochromis* species are beautiful and interesting fishes, too, but are out of place in this book because of their size and the not inconsiderable burrowing they sometimes do.

The males of the *Pseudocrenilabrus* species also have spots on their anal fins which the females are meant to, and do, mistake for eggs. Their yellowish or reddish terminal lobe serves this purpose. Being a true mouthbrooder, the female quite often takes the eggs just produced into her mouth before the father has a chance to fertilize them. Then the mother follows the male who is moving over the bottom layer discharging sperm. She gets very close to his genital region and not infrequently takes hold of the tip of his anal fin which in shape, size, and color looks very much like an egg (and an egg she would have to pick up!) At this point she gets spermatozoa into her mouth and fertilization of the eggs already gathered together there can take place!

By these mock-eggs the males are readily distinguishable from the females (which, in any case, remain noticeably smaller as a

A female *Pseudocrenilabrus* is seen here in the process of receiving sperm being discharged by the male. This is a spawning pair of *P. philander.*

rule). Another aid to differentiation are the male's much more magnificent colors. In this case, too, it is advisable to purchase more females than males wherever possible, even if the social structure of *Pseudocrenilabrus* species is quite different from that of the fish which otherwise prompt this advice. The mother carries the spawn in her mouth for a period just approaching two weeks and cannot feed during this time. Obviously this is a very strenuous task and the aquarist must make sure she is not exposed to the male's charms again too soon. If the mouthbrooding phases succeed one another two rapidly the female is quickly worn out. To have three females to one male would mean three females sharing the burden of motherhood between them.

PSEUDOCRENILABRUS MULTICOLOR
Dwarf Egyptian Mouthbreeder

This species is one of our oldest aquarium fishes. It also happens to be the smallest and at the same time the least demanding mouthbrooder. The males grow to a mere 8 cm in length. In its native range, all of eastern Africa from the lower Nile to Uganda and Mozambique, the species is not uncommon. The water temperature for keeping and breeding should again be 23 to 26°C.

A female dwarf Egyptian mouthbrooder that is obviously incubating eggs in her mouth.

The fishes can also, quite easily, be bred in the community tank. In the spawning season the male makes a pit in the sand by rapidly spinning round and round and beating his fins. Larger stones are pushed out of the way with his snout. Afterwards the male tries to lure a female into the spawning place. When the 30 to 100 eggs have been produced and are inside the female's throat pouch she should be protected from further advances by her ex-husband. Usually we find the

Pseudocrenilabrus multicolor, female in photo above and the male below. A female carrying a load of eggs shows an expanded gular area. The male is definitely more colorful than the female.

This female dwarf Egyptian mouthbrooder appears ready to incubate eggs.

female exhausted in a more sheltered corner of the aquarium (sheltered by floating plants, for example). Underwater we carefully drive her into a glass jar and with the jar transfer her into another tank. If we use a hand-net there is a danger that she might spit out the eggs that have just been picked up. If there is only one "couple" in the breeding tank, it is of course much simpler and safer to net out the male instead. After about 11 to 12 days the fry leave the mother's mouth for the first time. Only now may we offer food to the female again! At this stage the family can already be split up. It is riskier, but nicer, too, to continue observing the fishes for a few more days. During this time the little ones return to the maternal mouth if there is any danger, real or imaginary, and to spend the night. With *Artemia* they are easy to raise. The mother should be given a generous, nourishing diet for a month before being allowed near the male again.

PSEUDOCRENILABRUS PHILANDER

Dwarf Copper Mouthbreeder

This fish, which glitters golden or greenish depending on the light, occurs in almost all African states south of Angola, Zambia, and Mozambique. There, it is widespread in rivers and lakes and forms many subspecies. Some races grow to 12 cm,

others to no more than about 8 cm. The females are distinguished from *P. multicolor* females by their caudal fin, in *P. philander* females the lower half appearing yellow.

The keeping and breeding is the same as in *P. multicolor.* The two species should not be kept in the same tank as this would very probably result in the production of hybrids.

The genus *JULIDOCHROMIS*

These slender fish are rightly among the most sought-after of the dwarf cichlids. Some of them have such decorative colors that they can be compared to coral fish. A further asset is the fact that they are undemanding. They accept any food and have a preference for hard water (i.e. the kind of water that comes out of the tap in most households).

Members of this genus can be kept—and even bred—in community tanks of suitable design, that is with plenty of caves to hide in. More pleasure is derived from them, however, if they are accommodated in a rock aquarium such as I have described earlier and this tank is reserved for a single species. Slate tiles could, of course, be used instead of stones. A *Julidochromis* species tank with a length of 80 cm would be excellent, although the ideal would be a still larger size. Initially, a stock of one male and one female will do—in the course of a few months their offspring will bring

Julidochromis transcriptus, from the northern shore of Lake Tanganyika.

the population to the desired level.

The only problem we have with representatives of this genus is the selection of the first pair. Here, luck plays a major part. The sexes are not clearly distinguishable. If we put an individual into a tank with a fish that has already been there for some time we can be certain that the pair will quickly be locked in mortal combat. It is better to place both (one hopes) parents-to-be in the strange environment at the same time. Even then, however, we should be ready to apply the "dividing glass-pane method" I described under *Nanochromis* if necessary. Even relatively minor disturbances, the arrival of new plants, partial water changes, etc., can lead to the revival of hostility between two partners—including fish that have known each other for a long time. Caution, therefore, should be used when doing anything to this tank!

Facing page: A female *Julidochromis marlieri* cleaning the eggs (upper two photos). The same female guarding the nest from outside intruders (bottom photo).

In a well-established *Julidochromis* tank where things have settled down we can watch how generation after generation of these slender cichlids grow up in the parents' territory. When the fry have reached a certain size they leave the parental territory to their younger siblings and swim across to the "adolescents" in another part of the aquarium. In an 80-cm tank it is perfectly possible for two *Julidochromis ornatus* pairs to exist side by side without any problems.

The five species known so far are all native to Lake Tanganyika:

J. marlieri: dark brown body with 3 horizontal rows of whitish spots; northern end of the lake; 13 cm maximum.

J. transcriptus: upper half of the body black with 2 whitish, interrupted rows of spots, underside light in color; northern end of Lake Tanganyika; 7 cm maximum.

J. regani: body having four horizontal bands, the lowest one directly at the insert of the pectoral fins; widely distributed in Lake Tanganyika; 13 cm maximum.

J. ornatus: body having three horizontal bands and a large round black spot at the base of the caudal fin; northern and southern ends of Lake Tanganyika; 8.5 cm maximum.

J. dickfeldi: three horizontal bands on the body, no spot on the caudal base; southwestern

Julidochromis regani.

area of Lake Tanganyika; 8.5 cm maximum.

From a point of view of size, only *J. transcriptus, J. ornatus,* and *J. dickfeldi* can properly be included among the dwarf cichlids. In their natural environment these species also differ ecologically from their larger relatives. Whereas *J. marlieri* and *J. regani* live at depths of between 2 and 20 meters in typical "rock-scapes," the smaller species are said to favor a water depth within the 1 to 4 m range. There, their habitat is adjacent to sandy regions and has a bottom layer of small rocks, coarse gravel, and flat stones.

Photos in this page: *Julidochromis ornatus* is the most known member of the genus. An attractive and easy-to-breed fish, the "julie" is a popular aquarium fish the world over.

The color and markings described for *Julidochromis dickfeldi* appear distinctly in this photo. It is no problem to breed this fish.

JULIDOCHROMIS DICKFELDI
Brown Julidochromis

This species was not discovered until 1975 when a collecting trip was made to the southwestern part of Lake Tanganyika. It differs from the other *Julidochromis* species by its light brown basic color. The fins look very striking, all of them, with the exception of the pectoral fins, having a bright blue border. This species is no longer rare. As regards keeping and behavior, this fish is unlikely to differ from the other *Julidochromis* species.

ADDITIONAL SMALL CICHLIDS FROM LAKE TANGANYIKA

Lake Tanganyika is the home of a few more mini-cichlids that can be recommended unreservedly. These fishes have only very recently come to us. Like the *Julidochromis* species, they all demand medium-hard to hard water and a slightly alkaline pH (between 7 and 8.5) as well as plenty of caves to hide in.

Eretmodus cyanostictus, a yawning clown.

GOBY CICHLIDS

Of all Lake Tanganyika cichlids described here the quaint little goby cichlids present the greatest problems. Foodwise they are not choosy, but they demand very, very clean water and if we do not partially change their water at frequent intervals they will give us little pleasure. Aeration is a must where these fish are concerned and, if possible, a filter should also be employed.

These fishes are true bottom-dwellers. Usually they actually lie on the bottom-layer. With the mouth located

on the underside of the body they like to feed upon the algae on the stones. Free-swimming visibly tires them.

The difficulties we have with these fishes are particularly great when it comes to breeding. They are mouthbrooders but there is close pair-bonding. For mouthbrooders this is unusual. The act of spawning has been observed comparatively often. The fish spawn, quite unhurriedly, on a stone cleaned by them for that

Facing page: *Spathodus erythrodon.* Note the distinctive light blue spots on the head and body of this goby cichlid.

The Tanganyika clown has both light and dark vertical bands from the head to the base of the tail, quite distinct in this photo.

purpose. The eggs are immediately taken into the female's mouth. Unfortunately this is the stage at which most breeding attempts come to a sticky end.

At the time of writing the goby cichlids are still regarded as belonging to three different genera. The species that is most frequently imported is *Eretmodus cyanostictus,* the Tanganyika clown. The fish is brownish in color and has 8 or 9 narrow light cross-bands on the flanks. *Spathodus erythrodon* lacks these striations but instead has bright light blue speckles on the head and in the dorsal region. The third species imported into Germany, *Tanganicodus irsacae,* has both the cross-bands and the iridescent blue specks. As compared with the other goby cichlids, this species has a conspicuously pointed snout and a relatively small mouth.

TELMATOCHROMIS BIFRENATUS
Two-Banded Cichlid

This slender fish, which grows to approximately a mere six centimeters, is one of the smallest cichlids we know. The species is distinguished by a horizontal stripe extending along the base of the extended dorsal fin and another such stripe which runs from the tip of the snout through the eye to

Tanganicodus irsacae has both crossbands and blue spots.

the caudal base. The second stripe, that is, in the posterior half of the body, is traversed by 6 to 10 oblique lines. Keeping and breeding are more or less the same as for *Julidochromis.*

NEOLAMPROLOGUS LELEUPI

Lemon Cichlid

This species, which attains a length of about 10 cm, comes in two forms, the normal blackish one, *N. leleupi melas,* and the uniformly orange-yellow subspecies *N. leleupi leleupi.* The iris of these fish is blue.

There can be no doubt that the yellow form is the most popular one. It is found in Lake Tanganyika at depths of 15 to 20 m. The fish deposit their eggs (up to 300) on vertical walls. The "caves" provided for this purpose should have very broad entrances, not the narrow little holes preferred by *Nanochromis, Pelvicachromis,* etc.

The fry hatch out after about three days. After another six to eight days they are free-swimming. The care of the young, which search the bottom for food *(Artemia!),* is almost exclusively taken on by the mother. In water which is not too soft the raising of the fry presents no problems. The youngsters are dirty grayish brown in color, thus reminding one of the "fear coloration" displayed by the adult animals. Ideally, several individuals of this perfectly peaceful species should be kept together in medium-sized aquaria with plenty of caves and plants.

NEOLAMPROLOGUS BREVIS
Snail-Dwelling Dwarf Cichlid

Because of their droll behavior, Snail-dwellers are quite fantastic dwarf cichlids for the community tank. They need medium-hard or fairly hard water and neutral or slightly basic pH values. As very peaceful dwarfs that grow to a maximum length of 5.5 cm, they are ideal for housing together with small and medium-size rasboras, as well as with small barbs or live-bearers. In any case, however, fairly large snail shells must be added to the aquarium. The cichlids use them as living and spawning cavities, into which they withdraw at any suspected or actual danger—fascinating behavior for any observer!

A spawning setup appropriate for *Neolamprologus leleupi*; a cave-like environment is not necessary.

This page: It is not surprising why *Telmatochromis bifrenatus* deserves the two-banded adjective. **Facing page:** *Neolamprologus leleupi*, the more popular lemon yellow subspecies pictured here.

There are several species of Snail-dwellers. *Neolamprologus brevis* and *Neolamprologus ocellatus* are the most frequently offered. *Neolamprologus brevis* has a conspicuous light-blue stripe under the eye, and is characterized by eight to twelve light-blue iridescent vertical body stripes on an ochre-colored background. *Neolamprologus ocellatus* lacks these stripes.

Particularly beautiful is the only very recently introduced, at present still not scientifically named Snail-dweller, which has been given the name *Lamprologus "magarae."* Males grow to 7, and females to 5 cm in length. In the head region the fish have a blue marking that resembles that of *Neolamprologus brevis,* but, with the exception of a few bluish stripes in the back area, they have no other vertical body stripes. A distinguishing characteristic of this fish is the thread-like elongation on the top and bottom of the tail fin. The ventral, anal, and dorsal fins are also tapered.

Neolamprologus brevis, the rounded anal and caudal fins distinguishes this species from *L. "magarae."*

Neolamprologus elongatus is not choosy with respect to the selection of nesting site; a vertical wall, a flat surface or a cave can be used.

NEOLAMPROLOGUS ELONGATUS
Fairy Cichlid

This elegant and extremely popular Lake Tanganyika cichlid shall bring this little book to its conclusion. In the past it bore the name *Lamprologus savoryi elongatus* and was called "princess of Burundi," but the German "fairy cichlid" is much nicer. Although lively, the fish can be kept in the community tank. This cichlid also attains a size of about ten centimeters. The sexes can only be distinguished with difficulty. The male is very slightly smaller and has slightly more protracted fins.

This species is very easy to breed. The eggs (about 200) are laid in cracks and caves and guarded by the female. The fry hatch after 60 to 85 hours, and after another three days they are free-swimming. The parents do not molest their offspring at any stage. Similar to *Julidochromis* species we can allow successive broods to remain in the tank.

A true albino *Neolamprologus elongatus* is now bred commercially and seen in some pet shops.

The species, formerly designated as *Lamprologus brichardi,* is now classified as *Neolamprologus elongatus.* For a number of years a variety has been imported that is characterized by blue fin edges and a conspicuous blue marking in the head area: the Daffodil Snail-dweller or *Neolamprologus "daffodil."* At present it is uncertain whether it is a question of a still-to-be-described subspecies or only of a color variety of *Neolamprologus elongatus.*

Red color variety of *Apistogramma bitaeniata.*

Index